JOHN J. FAGAN

Department of Geology
City College of New York

The Earth Environment

PRENTICE-HALL, INC.
Englewood Cliffs, N. J.

Library of Congress Cataloging in Publication Data

FAGAN, JOHN J
The earth environment.

Includes bibliographies.
1. Earth Sciences. 2. Environmental protection.
I. Title
QE33.F32 333.7 73-10067
ISBN 0-13-222752-5

10 9 8 7 6 5 4 3 2 1

Printed in the United States of America

cc

PRENTICE-HALL INTERNATIONAL, INC., *London*
PRENTICE-HALL OF AUSTRALIA, PTY. LTD., *Sydney*
PRENTICE-HALL OF CANADA, LTD., *Toronto*
PRENTICE-HALL OF INDIA PRIVATE LIMITED, *New Delhi*
PRENTICE-HALL OF JAPAN, INC., *Tokyo*

The
Earth
Environment

To Celia and Henry Paley

Contents

2

WATER SUPPLY, 53

3

ENERGY, 89

4

OCEANS, 141

5

SOILS AND FOOD, 171

6

Preface

The quality of our future environment will require widespread understanding of the basic processes that underlie man's relationship to the resources of his planet. This book is an introduction to those earth processes and resources most closely involved in current environmental problems and crises.

To be a geologist, I have found, is both an advantage and a disadvantage in coming to grips with the seriousness of present day environmental problems. The advantage lies in familiarity with the slowness of formation of such earth resources such as fossil fuels and mineral resources and with the finite nature of such vital substances. The disadvantage lies in one's tendency to think in terms of great time intervals and very slow geologic changes, which too often sets up a barrier to the full acceptance of the seriousness of the very rapid changes now taking place as the result of man's new and vast power to overwhelm natural processes and exhaust the assets of the planet.

It becomes more obvious each day that the human species now wields an evergrowing power to damage its own environment on a truly planetary scale. Thus, it becomes increasingly vital that courses of study in earth science be made more relevant to current environmental problems and deal with the planet as the habitat of man, by relating the scientific understanding of planetary processes to the activities and needs of the human species.

I intend this book for use in either earth science courses or in interdisciplinary courses that deal with the interaction of man and his physical environment. A minimum of scientific and technical terms are used, and graphic and photographic illustrations are an essential part of the development of important concepts and processes.

In each chapter a major realm of the environment is considered, each beginning with selected basic principles and leading to the discussion of current environmental developments and problems within that realm.

After an introduction to the relationship between natural resources and human population, the first and second chapters of this book deal with the two most vital of earth resources, air and fresh water. The third chapter focuses on the modern uses of energy. The fourth chapter deals with the role of the ocean in human lives. The fifth chapter relates the world's soils to man's food supply. The last chapter considers the problems of supply and exhaustion of ore deposits and the proliferation of wastes caused by rapidly rising rates of consumption and population growth.

Suggested further readings are listed at the end of each chapter, and the general readings that follow this preface are highly recommended.

I am grateful to Constance Stallings for expert editorial assistance and encouragement at an early stage in the preparation of the book and for the able advice and criticism of Shirley Luban in the later stages.

JOHN J. FAGAN

General Readings

DASMANN, RAYMOND, *Environmental Conservation,* 2nd ed., John Wiley, New York, 1968.

EHRLICH, PAUL and ANNE H., *Population, Resources, Environment,* W. H. Freeman, San Francisco, 1970.

Environment (formerly *Scientist and Citizen*) A publication of the Scientists' Institute for Public Information, 438 N. Skinker Blvd., St. Louis, Mo., 63130. (Diverse, non-technical information.)

MCKENZIE, GARRY and RUSSEL UTGARD (eds.), *Man and His Physical Environment,* Burgess, Minneapolis, 1972.

MURDOCH, WILLIAM (ed.), *Environment: Resources, Pollution and Society,* Sinauer, Stanford, Conn., 1971.

REINOW, ROBERT and LEONA T., *Moment in the Sun,* Dial, New York, 1967.

Introduction

This book is about the clash of evergrowing numbers with a nongrowing habitat. It deals with the environmental problems caused by the runaway population growth on a planet that has unalterable dimensions but fixed or shrinking resources.

The evolution of man and all forms of life has been in response to environmental conditions. For most species, and for man prior to civilized living, these environmental factors have been rather simple ones such as the availability of food, water, and warmth. But now that man has developed a mechanized and electrified way of living and vastly increased his numbers, the picture is more complex. For modern man is a consumer of vast quantities of the earth's resources, a consumer whose appetite and needs have been significantly altering his environment and threatening his own future.

KINDS OF RESOURCES

The familiar term *natural resources* refers to supplies of water, food, soil, minerals, fuels, and even the air we breathe. Some resources, such as fresh water, can be replaced each year by the evaporation–rain cycle; these are the renewable resources, whereas others are *nonrenewable resources* which are consumed and lost by being used up.

Although production of renewable resources like food or water can be endless, there is always a limit to the rate at which they are produced. Consider the flow of a river (Fig. I.1). Each of the various parts of the

Figure I.1 The flow of rivers is part of the hydrologic cycle: water from rainfall returns to the sea. (Photo by author.)

earth's surface have a relatively fixed annual rainfall. Any one river has an annual flow limited by the rainfall of the region from which it flows. The river may flow constantly year after year but the *rate* of renewal is limited by the annual rainfall.

It is possible to deplete a renewable resource by using it faster than the renewal mechanism. In the case of the river, draining or pumping the water for irrigation purposes at a rate faster than natural replenishment can occur (through upstream rainfall), would deplete the resource. Heavy discharge of pollutants into the river may have the same practical effect in that the river is lost as a source of fresh water (Fig. I.2).

Figure I.2 Water pollution can eliminate our water supply resource. (USDA Soil Conservation Service.)

Figure I.3 Strip mining of coal in Ohio, an example of the use and destruction of a nonrenewable resource. (USDA Soil Conservation Service.)

The earth contains limited amounts of the nonrenewable resources, most of which can be used only once. Coal, for example, is formed by geologic processes so slow that a one-foot thickness of the fuel requires thousands of years of plant growth. Yet that foot is usually mined out in minutes (Fig. I.3). Iron ores, petroleum, and fertile topsoil are all nonrenewable resources in the sense that they are being formed by geological processes acting hundreds or thousands of times slower than the rates at which man is using (and depleting) them.

In many regions the availability of renewable resources is being outstripped by demands of growing numbers of consumers. Depletion and deterioration trends barely perceptible 20 or 30 years ago have become crises today and the heavy consumption of resources has caused pollution and other serious environmental problems, many of which did not exist or were not acute until recently.

NUMBERS OF PERSONS

The planet earth has recently been compared to a spaceship moving through the emptiness of space with all the oxygen, water, and other substances needed to maintain life. The analogy is a good one; in spite of lunar exploration and the investigation of other planets, there is no reason to believe that man will ever long survive outside of the planet on which he evolved. We have only this one home; its limits are our limits.

The earth is about 25,000 miles around but only about $\frac{1}{6}$ of it is habitable or very useful to man; the rest is covered by water or ice or consists of desert or rocky mountain slopes. This habitable sixth is already the home of over 3.8 billion (3,800,000,000) human beings, a constantly increasing number. Every nine seconds another person is born; every 16 or 17 seconds another dies. This means that many more humans are being added to the population than are dying; over 5500 extra humans each day! Before the year 2000, the world population is expected to reach six possibly seven billion persons.

The world is expected to double in population in a little less than 30 years. Every three years the number of humans increases by an amount equal to the size of the entire population of the United States. The incredible rate of increase of world population (Fig. I.4) presents us with a problem that is essentially a new one in human experience. The reality of the implications of such growth is difficult for most of us to grasp, but, clearly, the numbers of man must eventually reduce the capacity of the environment to supply his needs.

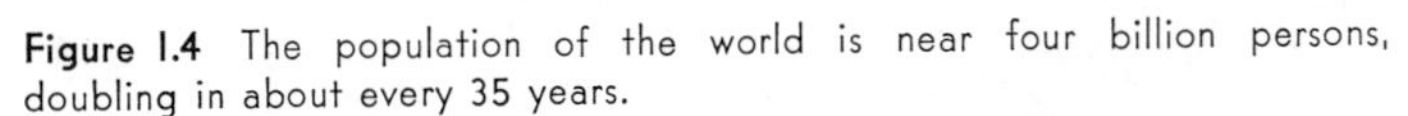
Figure I.4 The population of the world is near four billion persons, doubling in about every 35 years.

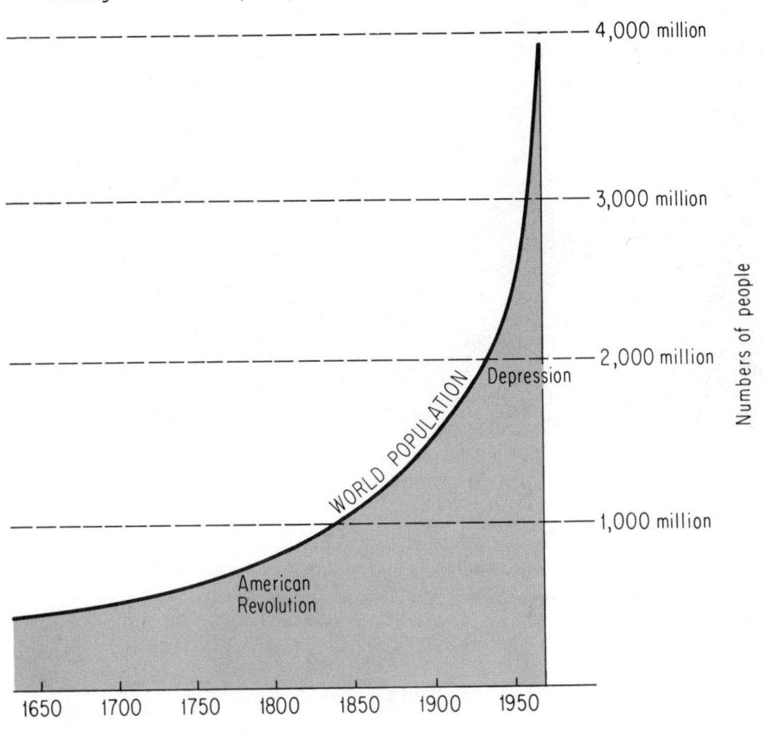

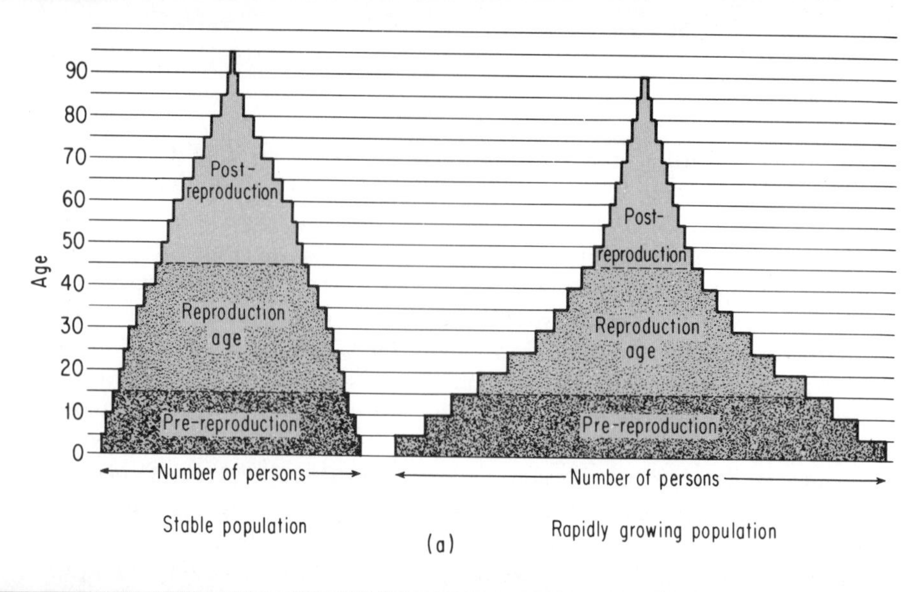

(b)

Figure 1.5 (a) A rapidly growing population differs from a stable population in its large proportion of young, reproductive age and pre-reproductive age persons. (b) Rising populations require the use of ever greater areas for housing, thereby lessening the space available for agriculture, recreation, and other vital uses. (USDA Soil Conservation Service.)

In many of the underdeveloped countries, population growth has far outstripped increases in food production. About $\frac{1}{5}$ of all deaths in the world are the result of malnutrition-related diseases or starvation. The average caloric intake of over half of the people in the world is less than the nutritional minimum required for normal growth and activity (according to U.S. Department of Agriculture figures).

The rapid population growth in many of the poorer countries of the world has reduced living to bare subsistence and has prevented the economic development that underlies the higher standard of living characteristic of the industrialized nations. Future years promise still greater increases in the population of most of the poorer countries because of the very large proportions of children yet to reach the age of reproduction (Fig. I.5).

Many environmental scientists believe that the runaway population growth is a kind of world cancer that erodes much of the advantages of modern civilization as it gnaws at the food, space, and resources of the planet.

In the chapters to follow, we will consider each of the principal kinds of natural resources, beginning with the air itself, concentrating on critical factors of the deterioration and limitation trends of these vital elements of our environment.

Suggested Readings

CLOUD, PRESTON, E. (ed.), *Resources and Man,* W. H. Freeman, San Francisco, 1969.

DAVIS, KINGSLEY, "Population," *Scientific American,* Vol. 209, No. 3, 1963.

EHRLICH, PAUL, *The Population Bomb,* Ballantine, New York, 1968.

Population Bulletin, "Population and Resources: the Coming Collision," Vol. 26, No. 2, The Population Reference Bureau, 1970.

SKINNER, BRIAN, *Earth Resources,* Prentice-Hall, Englewood Cliffs, N.J., 1969.

United Nations, Statistical Office. *Demographic Yearbook,* 1970.

1
Air

Surely no one on earth has even seen our planet as well as the Apollo astronauts as they orbited the planet en route to the moon. Nevertheless, looking down from their spacecraft, the Apollo 7 astronauts were unable to see southern California because of its pollution layer. Seen from spacecraft, airplanes, or mountaintops, most large American cities today appear covered in a gray-brownish haze, their own envelopes of air pollution (Fig. 1.1). Yet, seen closeup from a city street, the air seems deceptively normal; the slight discoloration is too familiar to be noticed.

It is only on those few days of the most severe pollution with its attendant widespread coughing and eye irritation that most city-dwellers are ever even aware of the envelope of contaminants in which they live and must breathe. But in the past few years, the contamination of the atmosphere has been increasing so noticeably that more and more Americans have been awakening to the damages of air pollution.

Air pollution is usually considered to be the contamination of the atmosphere with man-made gases or particles which render it injurious to humans, plant, or materials. The danger of air pollution can be increased by natural atmospheric variations. The seriousness of air pollution to life depends on the concentration of the contaminants and is often affected by winds, temperature changes, air pressure, and other natural atmospheric conditions.

Figure 1.1 (a) Air pollution over Salt Lake City. (U.S. Department of Interior, Bureau of Land Reclamation.) (b) Air pollution layer over Colorado. (Photo by D. Weiss.)

CANOPY OF GASES

The earth's atmosphere is an envelope of various gases surrounding the planet, extending thousands of miles above the surface, and grading imperceptibly into the near-emptiness of interplanetary space. From the surface upward to an altitude of about 60 to 75 miles the natural chemical composition of the atmosphere is quite uniform. It consists of a mixture of gases plus small amounts of water vapor and suspended particles.

Of the gases in the lower atmosphere, nitrogen comprises about 78 percent by volume; oxygen about 21 percent; and the gas argon about $\frac{9}{10}$ of one percent. Very small amounts of other gases also occur, such as neon, helium, krypton, and xenon. Carbon dioxide, although forming only about $\frac{33}{1000}$ of one percent of the air, is important as an absorber of heat and as an insulating blanket, helping to regulate air temperatures and essential for the growth of plant life. Clouds, which are composed of tiny water droplets or ice crystals, also characterize the lower atmosphere throughout the world.

Above a height of about 55 miles, the atmosphere consists of four gaseous layers, each of distinct composition (Fig. 1.2). Lowermost is the molecular nitrogen layer extending upward to about 125 miles. Above this lies an atomic oxygen layer. Between about 700 and 2200 miles lies a helium layer and above this lies a layer of atomic hydrogen. No definite upper limit can be set for this outermost part of the atmosphere, but at a height of about 6000 miles the density of the hydrogen gas drops to about the same as that found throughout interplanetary space. This arrangement of gases in the upper atmosphere is in order of weight; the lighter float above the heavier.

The force of gravity crowds together the molecules of the atmospheric gases progressively more densely from the upper limits of the atmosphere to sea level. The compression is caused by the weight of the gas layers above. The tendency of gases to be compressed is so great that the bulk of the earth's atmosphere, about $\frac{3}{4}$ of it, is contained in the lowermost six to seven miles, compressed by the weight of overlying gas. The density is greatest at sea level and decreases very rapidly upward.

Air pressure, which is usually measured by a barometer, is the measure of the weight of overlying air. It has an average value of 14.7 pounds per square inch at sea level. On top of Mount Everest ($5\frac{1}{2}$ miles high), air pressure is less than five pounds per square inch, and at 10 miles up it is less than two pounds per square inch.

One of the changes in the atmosphere we would encounter if we were to climb a mountain or rise in an airplane is a drop in temperature, averaging about $3\frac{1}{2}°$ F for each 1000 feet. The air becomes progressively cooler at this rate up to about six to eight miles, thereby dropping to about $-70°$ F at that level,

The term *troposphere* is applied to this lowermost layer of the atmosphere through which temperature tends to decrease upward. The term *stratosphere* is applied to the layer above in which the temperature is nearly constant (Fig. 1.2).

The troposphere contains almost all of the water vapor of the atmosphere and, therefore, nearly all clouds, storms, and precipitation. Jet airplanes commonly fly in the stratosphere in order to avoid weather uncertainties.

Figure 1.2 Diagrammatic cross section of the earth's atmosphere showing broad chemical subdivisions.

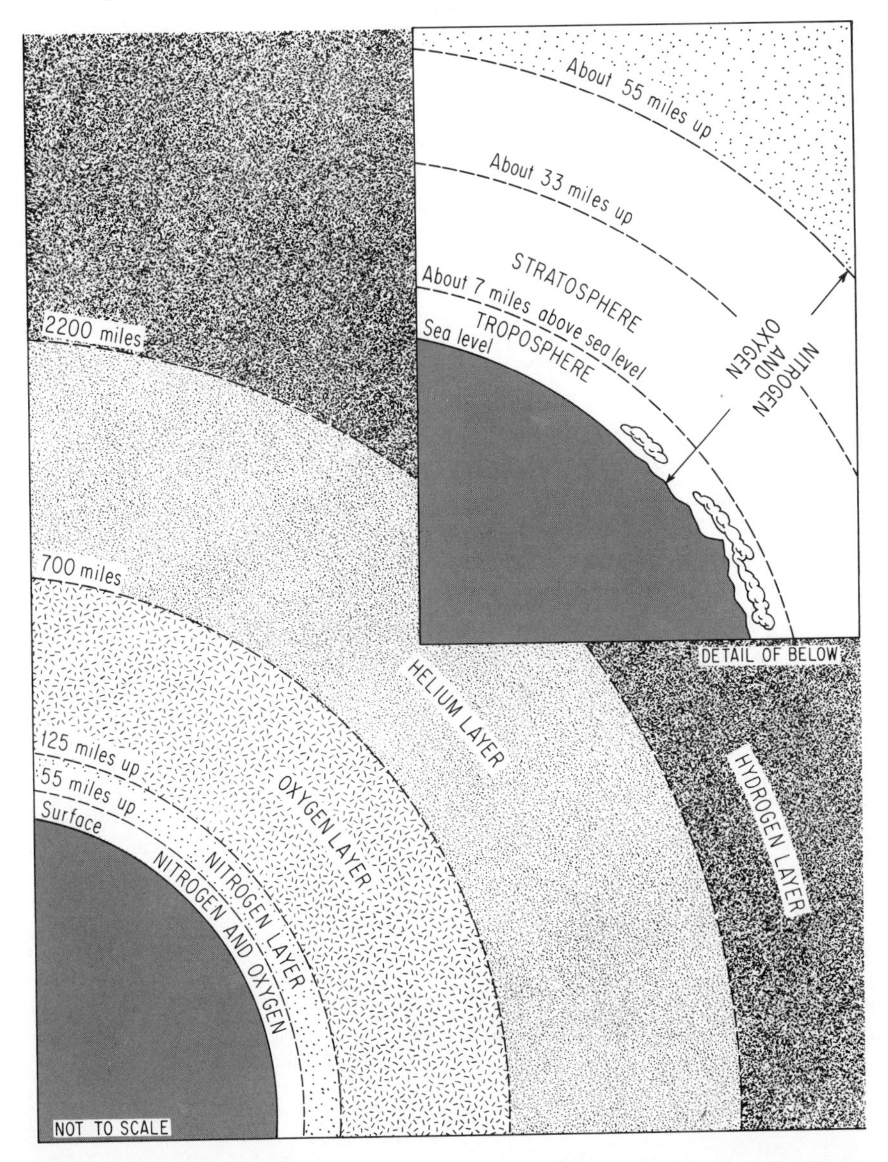

THE FLOW OF AIR

Convection and Winds

Wind is an air current or motion relative to the surface of the earth. Winds exist because of inequalities in temperature and pressure within masses of air which usually result from different rates of heating and cooling. A fundamental principle of air movement is the *convection* flow that may result wherever unequal heating of the atmosphere occurs. Strong solar heating of a part of the earth's surface causes the layer of air above it to expand and increase in depth and results in an outward flow of air (wind) toward cooler regions.

Simple convection air movements are commonly encountered along a coastline. On warm summer days rapid buildup of heat on the land surface adjacent to the sea causes the air over the land to expand, rising upward and outward (at high level) over the cooler sea from which air moves in along the surface onto the land; this is the sea breeze (Fig. 1.3). At night the land loses heat more rapidly than water by radiation, and

Figure 1.3 Winds caused by differential heating: (a) During the day, land surfaces become hotter than the adjacent sea; air over land expands and rises, permitting air over cooler sea to move onto land as sea breeze; (b) During the night, the reverse occurs as land loses heat faster than the sea, causing wind from land to sea.

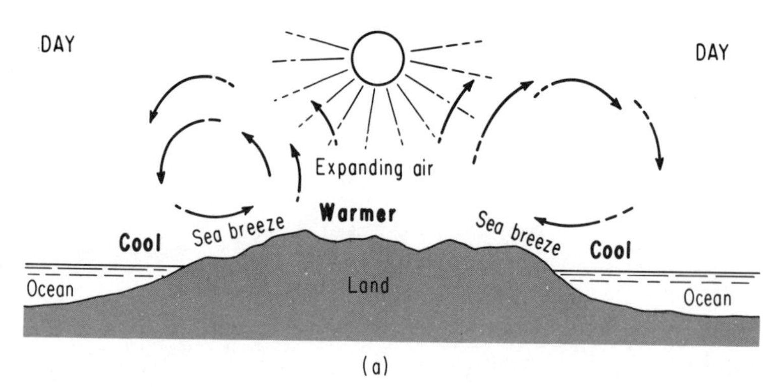

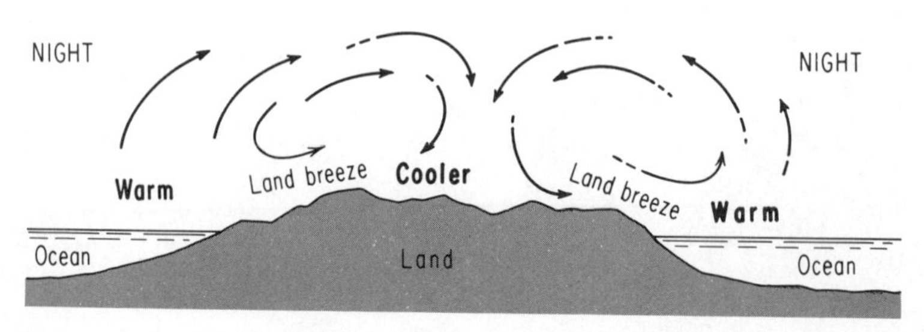

the air over the land is cooled more than that over the water. Now the conditions are reversed; the lower air layer over the sea expands, rising upward and out over the land at high level while, at low level a flow of air moves out from the land.

Planetary Circulation

The rate of heating over the planet as a whole is far from uniform. The greatest buildup of heat from solar radiation occurs near the equator, and the lowest occurs near the poles. A broad planetary circulation of air occurs as a result of this inequality based upon heat convection principles.

To visualize the manner in which air circulation develops, first imagine a nonrotating planet with no surface differences. Near the equator, the great buildup of heat causes air to expand and rise. The rising of air is

Figure 1.4 Theoretical flow of air in the atmosphere of a nonrotating planet, caused by expansion and outward flow of air at high altitudes in equatorial regions; return circulation at low altitudes causes wind toward equatorial regions.

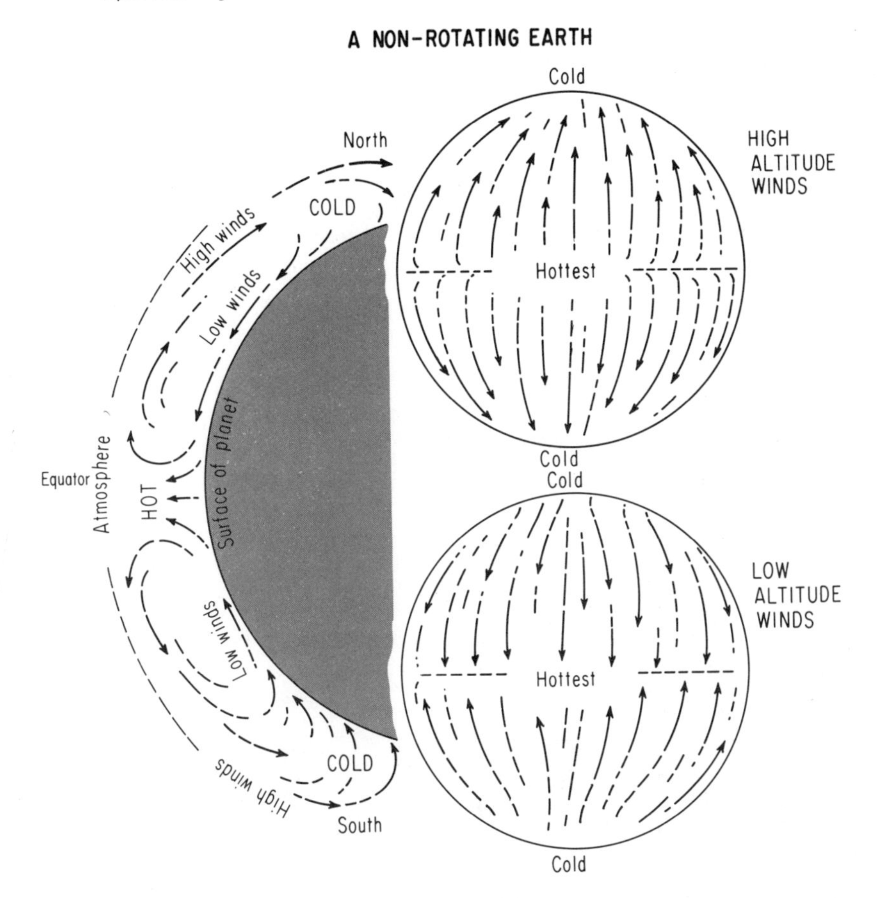

replaced by a "void," compensated for by air being transfered in from the cooler polar regions. In this simplified convection model, air flows poleward at high level and equatorward at low level; all motion of the air would be north-south in direction (Fig. 1.4).

As a next step in deducing the nature of atmospheric circulation, we add a rotary motion to our planet. This means the introduction of the *Coriolis force* which is produced by the effect of the rotating earth on air or any fluid moving on its surface. In the Northern Hemisphere the force acts to the right hand of the direction of motion, and in the Southern Hemisphere it acts toward the left hand. At high level, air over the equatorial belt begins to move poleward; it is deflected to the right (Northern Hemisphere) to become the system of upper westerly winds (from the west) at about the 30th parallel. In the Southern Hemisphere deflection to the left also turns the poleward flow at high level into a westerly wind system.

The upper air westerlies constitute great belts moving around the polar low pressure regions of the Northern and Southern Hemispheres (Fig. 1.5). The flow of the westerlies is not a simple circle but an endless succession of undulations or waves and eddies that include localized lows or highs of barometric pressure. Wind speeds vary from place to place and at different times. Maximum speeds of over 200 miles per hour are found in the jet stream.

But because the air moving poleward at high levels has been turned into westerly flows, the air tends to pile up at these latitudes more rapidly than it can escape poleward. This buildup takes place between 20 and 30 degrees North and South latitudes, producing at the surface two belts of high barometric pressure, one in each hemisphere. In these belts air

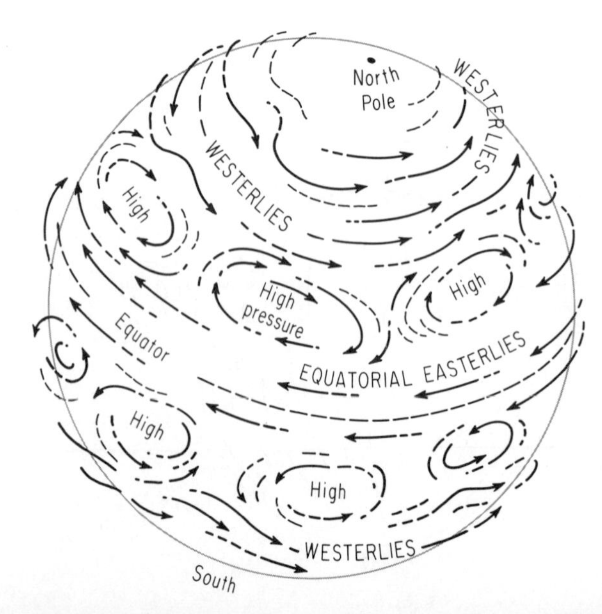

Figure 1.5 Generalized circulation pattern of world atmosphere resulting from the rotation of the planet and other factors in addition to differential heating from equator poles.

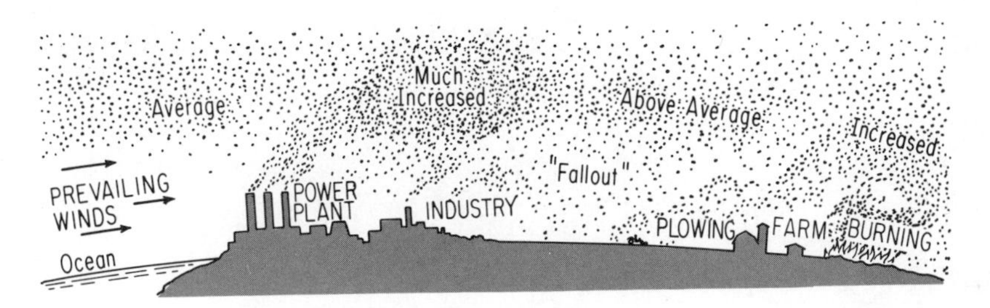

Figure 1.6 Particulate pollution distribution depends on prevailing winds caused by planetary circulation patterns and location of various pollution sources.

tends to subside, spreading out both equatorward and poleward, producing systems of prevailing surface winds.

The surface air that flows back toward the equator is turned westward by the Coriolis force to become the tropical easterlies (trade winds) well known to mariners. The tropical easterlies extend all around the earth over the equatorial regions, reaching high into the atmosphere, affecting movements of the oceans and transporting dust and other particles great distances.

The ceaseless flow of air over the surface of our globe has helped to spread the man-made air pollutants that have been increasing rapidly in recent decades (Fig. 1.6). But without the planetary circulation and prevailing winds, those pollutants would fail to be dispersed, tending to remain concentrated near their sources and quickly becoming lethal to inhabitants of cities and industrial regions. (See geography of pollution below).

AIR POLLUTANTS

With the growth of civilizations, man has been burning fossil fuels in ever-increasing quantities. Most air pollutants are the direct result of combustion. Most electricity ultimately derives from the burning of coal or oil. Thus, the most spotless of our electrical appliances cause the contamination of our atmosphere. Automobiles are moving "burners" spreading pollutants behind them as they run. In short most of the energy of civilization is supplied by burning, at the expense of the air we breathe. For a discussion of the actual present sources of energy see Chapter 3.

Among the effluent pollutant gases are carbon monoxide, sulfur dioxide, nitrogen dioxide, nitric oxide, ozone, vaporized hydrocarbons, and aldehydes. In addition, there are particulates, which may be dusts of plants or animal origin, many metals and metallic compounds, and insecticide particles.

Figure 1.7 Air pollution is growing rapidly in many regions. Winds distribute pollutants to regions far from the original source. (USDA Soil Conservation Service.)

The finest-sized particles are called *aerosols.* Unlike larger particles that settle rapidly, aerosols tend to remain suspended for long periods, sometimes forming visible fumes, smoke, and fine mists, and become invisible when widely scattered. Ordinary smoke usually contains minute particles, both solid and liquid, that are the result of incomplete combustion of fuels or waste (Fig. 1.7).

Unfortunately, the most dangerous atmospheric pollutants are invisible or nearly so, consisting of scattered aerosols or colorless gases such as sulfur dioxide and carbon monoxide.

Carbon Oxides

Carbon dioxide (CO_2) occurs naturally as a by-product of animal respiration and is an end product of complete combustion of fossil fuels or other carbon-containing substances.

A normal part of the natural growth-decay cycle of life, it is not ordinarily considered to be a pollutant or an inherently dangerous gas. But the enormous increase in combustion of fossil fuels in recent decades has caused a slow but measurable buildup of carbon dioxide in the atmosphere. The possible long-term effects of this concentration on the atmosphere will be considered later in this chapter.

Whereas the complete combustion (oxidation) of fuels or other carbon-containing substances results in the production of carbon dioxide, incomplete combustion produces carbon monoxide (CO), an almost exclusively man-made, toxic pollutant.

Unfortunately, carbon monoxide is invisible, odorless, and tasteless. Nor does it tend to noticeably irritate the eyes, nasal passages, or lungs. Instead, it passes through the walls of the lungs into the blood where it combines with hemoglobin to form carboxyhemoglobin, which has the effect of reducing the oxygen carrying capacity of the blood. A small amount of carbon monoxide in the air of the lungs can have a major effect on the oxygen transport function of the blood. All tissues of the body may suffer from oxygen deprivation, but the two tissues most sensitive to lack of oxygen are the brain and the heart.

The acute effects of carbon monoxide poisoning range from slowed perception and reaction responses to unconsciousness and death. The chronic effects are much less readily detected. Carbon monoxide exposure has been reported to be associated with heart disease. Many scientists now feel that the well-documented correlation between smoking and heart disease may be due in part to the carbon monoxide in cigarette smoke. It has also been suggested that the smaller babies produced by women who smoke frequently during pregnancy may be the effect of the carbon monoxide in the tobacco smoke.

The primary source of carbon monoxide in the atmosphere is automobile exhaust in which it is the result of the incomplete combustion of gasoline. An automobile, moving slowly along in heavy traffic, commonly builds up an interior concentration of 25 to 50 parts per million (Fig. 1.8). Long exposures to the gas in concentrations of over 50 ppm

Figure 1.8 Automobiles comprise the greatest source of air pollution in the United States. Most of the pollutants are invisible; most are hazardous to health. (Photo by author.)

are considered dangerous by most medical authorities. Cardiac conditions are known to be aggravated by exposure to auto exhaust fumes. A concentration of 1000 ppm of the gas can produce unconsciousness in an hour and death in about 4 hours.

Sulfur Oxides: From the standpoint of their widespread harmful effects on man and the problems involved in avoiding their production, the oxides of sulfur may be the most dangerous single pollutant. High sulfur oxide concentrations have been part of most of the fatal air pollution episodes that have occurred in London and other cities.

Sulfur dioxide (SO_2) is produced when sulfur-containing fuels are burned. Sulfur is found in coal and fuel oil in varying amounts so that combustion of these substances for heat or power always produces some sulfur dioxide. Sulfur trioxide (SO_3) is produced in the sunlit atmosphere by the oxidation of sulfur dioxide; some amounts may also be formed during combustion of fuels. Moisture in the air reacts rapidly with sulfur trioxide to form a mist of sulfuric acid (H_2SO_4).

Sulfuric acid is a very corrosive acid that damages living tissue, nylon stockings, and other substances, including some rocks. Sulfuric acid mist consists of tiny droplets of just the right size to be readily inhaled, with corrosive effects on lung tissue. Another sulfur-containing gas is hydrogen sulfide (H_2S), recognized by its odor of rotten eggs. It is even more poisonous to life than carbon monoxide and causes considerable damage to materials but is not as widespread a pollutant as sulfur dioxide. Hydrogen sulfide is usually associated with sewage or some industrial source.

Nitrogen Oxides

Oxides of nitrogen found in the air as pollutants are nitrogen oxide (NO) and nitrogen dioxide (NO_2). Both may be produced by hot combustion processes because of the chemical combination of atmospheric nitrogen and oxygen that occurs whenever high temperatures occur. Soon after its formation most nitrogen oxide is converted into nitrogen dioxide, a brownish gas with an unpleasant odor.

The effects of nitrogen dioxide on man depend on the concentrations of the gas; they range from mild irritation to severe lung damage. Commonly, this gas produces chronic respiratory ailments. Of great significance is the fact that nitrogen dioxide strongly absorbs ultraviolet light from the sun, creating nitrogen oxide and atomic oxygen (O). This oxygen then combines with the atmospheric oxygen molecules (O_2) to form ozone (O_3), a highly reactive substance implicated in the dangerous smog episodes in Los Angeles.

Hydrocarbons

These are gases (or aerosols) introduced into the atmosphere by evaporation of gasolines and other organic liquids, combustion of fossil fuels, or other means. The effects on man of particulate hydrocarbons are quite varied. Some of these substances are carcinogenic (cancer-inducing); some are irritating; some undergo further chemical changes in the atmosphere to produce other dangerous pollutants as considered below.

Smog Chemicals

Photochemical smog is produced in the air by sunlight acting on the oxides of nitrogen and hydrocarbons.

In the presence of sunlight, nitrogen dioxide combines with hydrocarbons to form oxidants, such as peroxyacetyl nitrate (PAN). The dangerous components in a typical photochemical smog are ozone, nitrogen dioxide, PAN, and other oxidants. Smog is rapidly becoming a common part of urban air, especially in cities with sunny climates and heavy automobile traffic.

Particulates

Particulates are nongaseous substances suspended in the atmosphere. They may be solids or liquids, come in a wide range of sizes, and are the main cause of visible air pollution.

Particle size controls the speed of settling. Very small particles settle so slowly that they persist in the air for long periods of time and may be carried long distances. Dust and aerosol levels are generally rising in our atmosphere; they are the products of all kinds of human activity, especially combustion by industry, power plants, incinerators burning in open dumps, and forest fires (Fig. 1.9). These particulates are added to natural products of volcanic eruption and wind erosion of soil, the latter aggravated by overgrazing and other man-caused damage (see Chap. 5).

Particles less than five microns in size are especially dangerous to life because they may reach the lower respiratory passages and lodge in the air sacs of the lungs. Sulfur dioxide is often absorbed on such tiny particles and slowly released into the air sacs. This is an example of a dangerous gas combined with particulate pollution.

Lead: This is one of the most potentially dangerous pollutants in our environment. Well over 500 million pounds of tetraethyl lead that were to be mixed with gasoline were produced in this country in 1971. During combustion of the gasoline in engines, about 70 to 80 percent of this lead is exhausted into the atmosphere in the form of aerosol- or dust-sized particles of toxic lead compounds.

a)

b)

Figure 1.9 (a) Burning of waste dumps adds massive amounts of pollutants to the atmosphere. (USDA Soil Conservation Service.) (b) Industrial complexes like this one in Tennessee are major polluters of the atmosphere. (USDA Soil Conservation Service.)

Lead has been used in gasoline since 1923, and its danger was quickly recognized by a few scientists but ignored by the public and by oil companies for over 40 years. Recent studies show that snow that has fallen in Greenland since 1940 contains up to 500 times more lead than did the snows that fell in pre-Christian times.

Lead enters the body via lungs, intestinal tract, and even the skin. In sufficient quantity, it can damage the brain, gastrointestinal tract, and other parts of the body. In the amounts inhaled and otherwise taken in by the average urban American, lead interferes with the manufacture of hemoglobin. Recent evidence indicates that only slightly higher levels of lead than those present in the *average* American city-dweller can cause mild anemia.

City children are especially susceptible to exposure to atmospheric lead because of their common exposure to lead-based paints often found in old city houses. It is possible that millions of urban and suburban children may already be mentally impaired as a result of the regular intake of lead emitted by automobile traffic plus the amount of lead obtained from leaded paints.

Asbestos: Asbestos is a fibrous mineral long used for insulation and fireproofing. In the last two decades asbestos has been commonly sprayed onto metal and other surfaces in buildings, replacing the much costlier method of enclosing supporting members in concrete. But in the spraying process as much as 50 percent of the asbestos escapes into the air, and the resultant aerosols comprise a growing pollution danger.

In a study made during the 1960s asbestos particles were recorded as being present in lung tissue of about *half* of the total consecutive autopsies performed in New York City. Heavy levels of exposure to these tiny, needle-like fibers have been linked to a greatly-increased risk of developing lung cancer.

Mercury: A number of metals other than lead have been found as particulate pollutants, especially in the vicinity of metallurgical operations. None of these metals seem to be widespread enough in the atmosphere to present a major danger to man, with the possible exception of mercury. Pollution of sea water by mercury is considered in Chapter 4.

It now seems that some of the oceanic mercury may originate from the smokestacks of power plants and incinerators. The mercury vapor from power plants is probably produced by the burning of coal containing minute amounts of mercury; that emitted by incinerators probably results from the fact that paper is a major part of municipal refuse and that mercury is often a key substance in the production of paper.

Mercury is well-known as a toxic substance when ingested in sufficient amounts but the health effects of low levels of mercury in the atmosphere are yet to be established.

Chlorine-Containing Particulates: Another pollution hazard recently recognized consists of droplets of hydrochloric acid released when the plastic called polyvinyl chloride (PVC) is burned in incinerators. Over a billion pounds of the plastic are produced each year in this country and much of it is eventually incinerated. The resulting hydrochloric acid becomes an aerosol or mist that can affect the skin or damage the upper respiratory system if inhaled.

DDT is the most famous of a number of man-made *chlorinated hydrocarbons* used as pesticides. It is not uncommon for half of the amount of a pesticide being sprayed on a field to remain suspended in the air as aerosols or larger particles. In this manner, toxic pollutants may be wafted great distances from the original farmland to be inhaled by a variety of creatures, including man. Eventually the pollutants may settle into the ocean or onto the land surface from which they are commonly washed out to sea, there to be concentrated by organisms of the marine food chain (Fig. 4.8). See Chapter 4 for a fuller discussion of the travels and dangers of this long-lived pollutant as it accumulates in the sea.

Radioactive Particles: Radioactive substances that spread through the atmosphere can originate from the exhausts of nuclear power plants, nuclear fuel processing plants, and from the testing of atomic weapons. The eventual settling of these substances from the atmosphere comprises the radioactive "fallout" danger, much discussed in the last two decades.

Radioactive pollutants include particles of short-lived isotopes, such as Iodine 131 with an eight-day half life; a number of isotopes with longer durations, such as Cesium 137 that has a half life of 30 years; and others with half lives measured in thousands of years.

The recognition of the danger of radioactive discharges into the atmosphere by nuclear power plants dates back to the studies in the 1950s when leukemia occurrences in children seemed to be caused by radiation damage from X-rays to which embryos had been exposed during examination of the pregnant mothers. This led to studies of birth defects and of leukemia in children which had increased in many parts of the United States in the aftermath of nuclear weapons testing and the fallout of radioactive particles.

Most of the nuclear reactors in operation during the last decade have gone through a history of gradually-increasing emissions of radioactive iodine, cesium, strontium, or other isotopes into water and air. The rise frequently results from the progressive deterioration of fuel rods during the operation of the plants.

For example, the Dresden reactor, which is located about 50 miles from Chicago, has been producing electricity since 1960. When it began operation, radioactive discharges into the atmosphere were very low but rose to 284,000 curies in the year 1962, to 521,000 curies in 1964, and to 736,000 curies in 1966. Infant mortality rates also rose in the mid-1960s in regions of Illinois located downwind from the Dresden nuclear power plant.

It now appears that the radioactive aerosols of gases released from reactor exhausts and other sources, once regarded as relatively harmless because of the small amounts and their "dilution" in the atmosphere, have a far more serious effect than ever anticipated (see Chap. 3). They may cause radiation damages from without or within the body because they may be inhaled into the lungs and may dissolve into the bloodstream where they continue to emit radiation.

Some scientists doubt that there is any "safe level" or threshold above which levels of radiation must rise in order for any damage to living cells to occur. They feel that *any* radiation to which a population is exposed becomes responsible for damage to children, adults, and even the unborn, in proportion to the intensity of the radiation.

SOURCES AND EFFECTS OF POLLUTION

Creating Pollution

In 1972 the United States produced over 140 million tons of air pollutants. Of this, about 60 percent came from automotive vehicles, about

Figure 1.10 Major air pollutants and principle sources. Box indicates solid particles; "cloud" indicates gases.

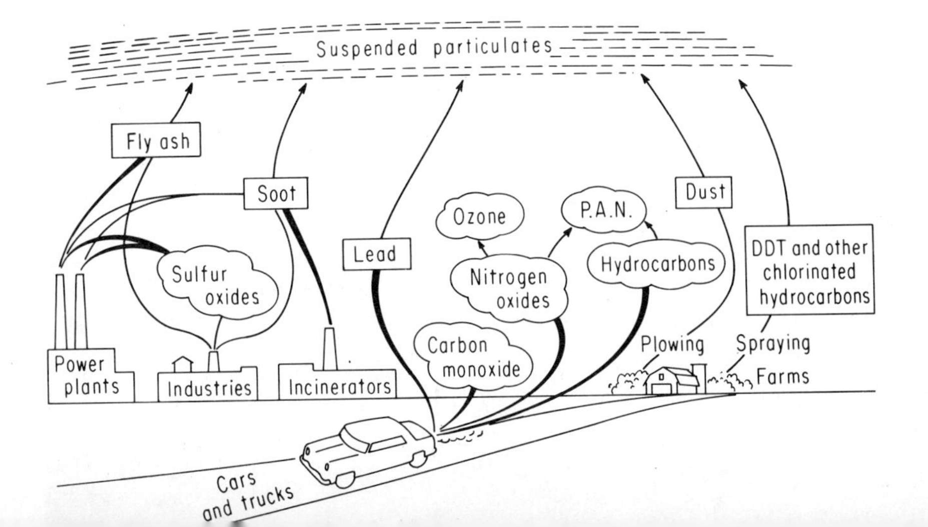

14 percent from power generation, about 18 percent from industrial plants, and about 8 percent from heating and refuse burning (see Fig. 1.10).

Automotive Sources: Automobiles and trucks use internal combustion engines that have built-in, perhaps insoluble, pollution problems (Fig. 1.11). The construction of the engine and its inherent inefficiency in fuel oxidation result in the production of large amounts of carbon monoxide, hydrocarbons, and nitrogen dioxide. In order to achieve quick starts and rapid acceleration, a high compression ratio is used which requires a fuel that usually burns with low efficiency, allowing hydrocarbons to escape through the exhaust and carburetor vents–especially during low speed driving.

Proposals have been made to return to steam engines for automobiles. Steam-propelled cars use external combustion engines which can produce considerably less carbon monoxide and hydrocarbons. But existing steam-engine designs are characterized by large size, greater cost than internal combustion engines, and other limitations.

Considerable research has gone into the development of propulsion by electricity-generating fuel cells, devices in which fuel could be completely oxidized, the byproducts being carbon dioxide and water. But such fuel cell-powered cars are far from realization. The use of cars powered by conventionally-produced electricity have advantages for use in heavily-polluted cities, but they must be charged with electricity produced in

Figure 1.11 Automobiles and trucks using internal combustion engines emit quantities of air pollutants. (Photo by author.)

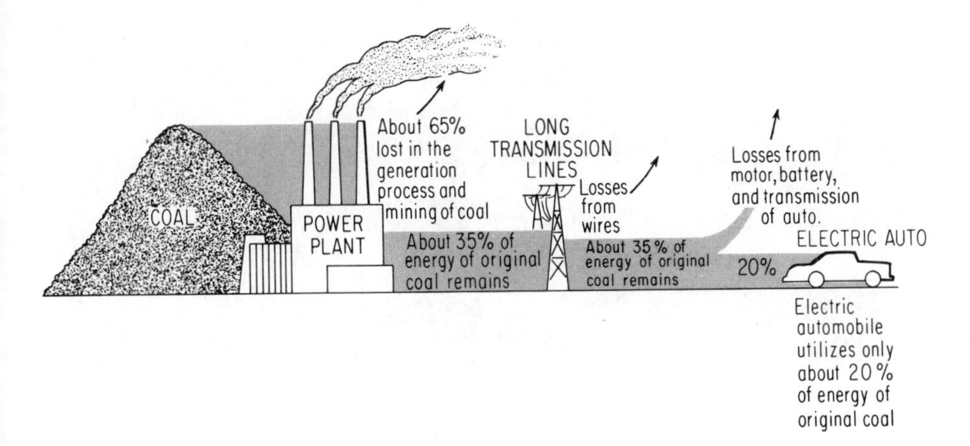

Figure 1.12 Vehicles run by electrical motors require the generation of electricity, usually by burning coal or other pollution-producing fuels. Such vehicles are themselves free of air pollutants, but their efficiency is so low as to represent the production of large amounts of pollutants.

large power plants. (Fig. 1.12). This would mean replacing the air pollution from gasoline-powered automobiles with the pollutants emitted by larger or more numerous power plants (see Chap. 3).

Power Production: Electric power production has increased over 650 percent in the last 25 years! Manufactured chiefly by burning fossil fuels, it is the fastest growing cause of air pollution (Fig. 1.13). The specific pollutants produced by power plants, in order of abundance are: sulfur

Figure 1.13 One of the new power plants in the Four-Corners region; this coal-burning electric-generating installation located in New Mexico produces vast quantities of electricity and air pollutants. (Photo by author.)

Figure 1.14 Sulfurous gases from this copper smelter in Tennessee have destroyed vegetation throughout this region. (USDA Forest Service.)

dioxide, suspended particles (especially fly ash), nitrogen oxides, hydrocarbons, and carbon monoxide. Actual ratios vary with the specific power plant and the fuel it burns.

The production of electric power, trends in terms of power sources and locations, are considered in Chapter 3.

Industry: A major source of air pollution is the operation of the hundreds of thousands of factories and industrial plants located throughout the nation. The proportions of pollutants produced varies with the specific industry. The most abundant on a nation-wide basis is sulfur dioxide, but particulates of various kinds are almost as common and include a great variety of substances, including mercury, lead, and other harmful metals, especially where the industry is an ore-smelting operation (Fig. 1.14).

Space Heating: Most homes in the United States are heated by fuel oil, natural gas, electricity, or coal, in that order of importance. Electricity

Figure 1.15 Home heating is a major source of air pollution like that found in New York City. (Photo by Jeremy Luban.)

for space heating produces air pollutants where it is generated, *not* in the building where it is used, so that such pollution comes under the heading of power production. Pollutants produced by the burning of fuel for space heating, in order of abundance, are: sulfur oxides, carbon monoxide, nitrogen oxides, hydrocarbons, and particulates (Fig. 1.15).

The air pollution produced in buildings from space heating has been decreasing in relation to pollution produced from other sources because of better home insulation and the trend toward heating by electricity. But this, of course, means still more pollution from the power plants where the electricity is generated (see Chap. 3).

Damage From the Air

It is only in the past few decades that Americans have really begun to recognize the extent, danger, and complexity of air pollution, and the lack of knowledge concerning the phenomenon. We may classify the effects of air pollution into four categories: (1) reduction of visibility and similar visual effects; (2) deterioration of materials; (3) damage to vegetation; (4) effects on man.

Visibility: The first noticeable effect of air pollution is often a decrease of visibility (see Fig. 1.16). Almost everyone who has approached a large city from the air has noted the gray pall of pollution, especially well-developed if there is no wind blowing. It is quite common for airplanes landing in large cities on cloudless days to have difficulty locating the airports visually. Pilots landing during "clear weather" in Phoenix, Arizona, a city once noted for pure air and azure blue skies, have had to rely on instruments in order to find the runway.

In some American cities, the effect of air pollution on visibility has been severe enough on occasion to curtail the flow of automobile traffic. The reduction of visibility is caused by the scattering of light by suspended particles in the air; the most intense reduction is produced by the larger aerosols.

Deterioration of Materials: Air pollutants can damage almost all solid materials including the hardest metals and stones. Acidic pollutants, such as sulfuric acid formed from sulfur oxides, are responsible for the corrosion of various metals and the disintegration of nylon and other textiles. Hydrogen sulfide gas blackens paints and tarnishes silver. Ozone is one of the pollutants that causes the cracking and deterioration of rubber.

There are numerous examples of the effect of the modern atmosphere on ancient stone art treasures. Statues and carvings made of marble are

(a)

(b)

Figure 1.16 Air quality can vary from day to day in the same region. (a) New York City on a clear day of good air ventilation. (b) New York City from the same point of view on a day of poor ventilation and heavy pollution. (Photos by author.)

especially susceptible. After surviving all the previous centuries the frieze of the Parthenon in Athens and many of the sculptures of Rome and Florence are all crumbling—some unrecognizable after decades of attack by air pollutants of today's cities. Even granite, one of the most geologically durable of rocks, yields to the corrosion of severe air pollution. A good example of this is "Cleopatra's Needle," an Egyptian obelisk transplanted to New York City. Its hieroglyphics had survived several thousand years but were finally largely obliterated by the city's atmosphere (Fig. 1.17).

Particulate pollutants driven at high speeds by the wind also cause damage to building surfaces. The total cost in the United States of these effects is very difficult to establish but has been estimated at several billion dollars a year.

(a) **Figure 1.17** Cleopatra's Needle, an Egyptian obelisk located in New York City, has undergone intense deterioration from air pollutant attack. (a) The obelisk soon after being brought to New York, thousands of years have failed to obliterate the carved hieroglyphics. (Courtesy of Metropolitan Museum of Art and New York City Department of Parks.) (b) The obelisk after a few decades of exposure to polluted city air. (Photo by author.) (b)

Damage to Vegetation: It has been estimated that the United States loses about ten million dollars worth of vegetables each year because of air pollution damage. But the full cost of such damage to crops, trees, and other plant life can probably never be assessed because of the subtle nature of the deterioration. In some instances, orchards have yielded 40 percent more fruit when the trees were protected by plastic and filters from the surrounding polluted air.

The most dramatic instances of pollution effects on vegetation have been seen in the total destruction of trees and other plant life in the areas surrounding copper smelters where sulfur oxides are produced by the "roasting" of copper sulfide ores (Fig. 1.14).

We now know that there are a number of ways in which plants can be damaged, and several gaseous pollutants have been held responsible. Ozone, oxidants like PAN, hydrogen fluoride, and nitrogen dioxide are among the gases that have been found to damage vegetation.

Ozone in the polluted air rising from the densely populated valley region east of Los Angeles had destroyed more than 1000 acres of majestic and valuable ponderosa pines in the San Bernadino National Forest by 1971 (Fig. 1.18). The forest is some 5000 feet above sea level but only 80 miles east of Los Angeles and its suburbs. Similar effects of ozone and sulfur dioxide on sensitive pine trees have been observed along large

Figure 1.18 Smog of air pollutants creeping up from Los Angeles and the San Bernadino Valley onto the 5000 foot mountains to the east. (USDA Forest Service.)

Figure 1.19 Pine trees on left grown in polluted air; same species on right grown in relatively unpolluted air. (USDA Forest Service.)

highways and in cities across the nation (Fig. 1.19). These and many other trees and shrubs die, become stunted, or undergo browning and malformation of leaves. New Jersey, a state heavy with refineries and other industry, and known to travelers for its frequent odorous and irritating air, has had virtually every crop damaged by pollution.

Even the simpler plants can be severely effected by air pollution. Lichens, for example, have disappeared from within many miles of the cities with the worst air pollution like New York and Los Angeles.

Effects on Man: (1) *Respiratory ailments.* There is consistent statistical evidence from the major metropolitan areas of the United States that heavy concentrations of air pollution are associated with ill health and an increasing death rate from certain chronic diseases including heart and lung diseases and various forms of cancer.

It is clear that the common disorders of the bronchial tubes and lungs called chronic bronchitis and emphysema or bronchitis–emphysema is showing alarming increase in some places in the United States. Chronic bronchitis and its complications is already the leading cause of death in men over 45 in Great Britain. Mortality from this disease has been correlated with the population size of cities and even with the amount of fuel burned in the cities.

The aggravation of chronic bronchitis–emphysema by air pollution has been most clearly demonstrated in this country by the Donora "disaster." Donora is a town in Pennsylvania located on a bend of the Monongahela River and surrounded by hills. The air above Donora tends to fill with smoke and fumes from the blast furnaces, steel mills, and sulfuric acid plants located along the river. In October of 1948, a warm front had moved into the Donora region, trapping cold river air in the valley. In this lower air sulfur dioxide and other pollutants emitted by the smokestacks became more and more concentrated. For three days Donora's pollution increased, the air became increasingly darker and less breathable. 20 persons died, 17 on the third day. Then a heavy rain fell, cleaning the air and ending the epidemic.

5910 persons were reported ill during the Donora episode. More than 60 percent were persons 65 years and older and almost half of these were seriously ill. In the first nine years following the episode, those who had become ill and recovered showed a higher mortality and incidence of illness than those who were present but unaffected, probably reflecting the long-term effect of the pollutants on those with prior damage to their lungs and hearts.

The incidence of emphysema in this country has been rising rapidly. During a single decade the death rate from that disease in California rose 400 percent. Some of this increase may represent better medical diagnosis, but it also indicates a rapid rise in a disease related to the worsening air pollution levels in the state.

It should be noted that all chronic respiratory diseases involve stress on the heart because the heart must work harder to compensate for the lack of oxygen absorption. This may partly account for the higher incidence of heart diseases among residents of large cities than among those who live in the country or in smaller, less air-polluted cities.

(2) *Carcinogenic and mutagenic pollutants.* There is strong evidence that some hydrocarbon air pollutants (such as benzopyrene) and some particulates are carcinogenic (cancer-producing). It is probable that benzopyrene, which is a common ingredient of polluted urban air, can also be a mutagen capable of causing mutation or deterioration of genetic material (DNA) in body cells. The seriousness of the threat to life presented by various concentrations of potentially carcinogenic and mutagenic air pollutants has yet to be determined.

(3) *Cumulative damage.* An important concern of many scientists who deal with air pollution and other aspects of environmental deterioration is the question of *cumulative* damage to life by repeated exposure to very small amounts of pollutant substances.

If substances, like sulfur dioxide, present in low concentrations, can be highly injurious to certain particularly susceptible persons after only a few days exposure, as in the Donora episode, how can we know that two or three decades of intermittent exposure to similarly low doses will not be injurious or fatal to many more people? This question arises again and again as we deal with the various toxic chemicals, the radioactive isotopes and potentially carcinogenic or mutagenic substances, all generally on the increase in the atmosphere and our environment as a whole.

Air Masses

Origin of Masses: The troposphere varies greatly from one region to another in its temperature and its moisture content (humidity). The lower troposphere is a mosaic of distinct masses of air which move about, causing the weather changes familiar to all of us. After developing properties from their specific regions of origin, air masses are then jostled along by the earth's prevailing circulation trends, outlined above.

Air masses usually extend a few miles up and have distinct boundaries with properties different from those of adjacent masses. A boundary between two masses is called a front. Most precipitation occurs along fronts rather than deep within the boundaries of an air mass.

An air mass originates when a portion of the troposphere remains relatively stationary for a time over a particular area of land or ocean and gradually takes on the properties of this region. For example, a cold, snow-covered region develops an air mass of low temperature, reflecting this source region. Similarly, the air over a warm, tropical ocean will be heated by radiation and tend to absorb water vapor by evaporation from the sea surface, resulting in a warm, humid air mass.

Air mass characteristics and directions of travel are critical in atmospheric studies because entirely different weather conditions may exist, depending on whether the air mass is warmer or colder than the surface over which it is passing.

North American Masses: The major air masses that control the weather of the United States do not originate here but move into the country as invasions from source regions on adjacent areas of sea and land. Figure 1.20 shows the major source regions and directions of movements of the air masses that regularly affect the nation's weather.

The polar and arctic continental air masses originate in a vast source area comprising the Arctic Ocean and northern Canada and Alaska. They move into the United States during the winter, passing down the Central Plains into the Mississippi Valley and across the Great Lakes region bringing cold weather to the nation. Crossing the Great Lakes the continental polar air masses may pick up moisture from the relatively warm lake surfaces, tending to become unstable and dumping snow as they reach the eastern and southern shores.

Another major source region is the North Pacific Ocean from which come cool, moist maritime polar air masses. They bring heavy rains to

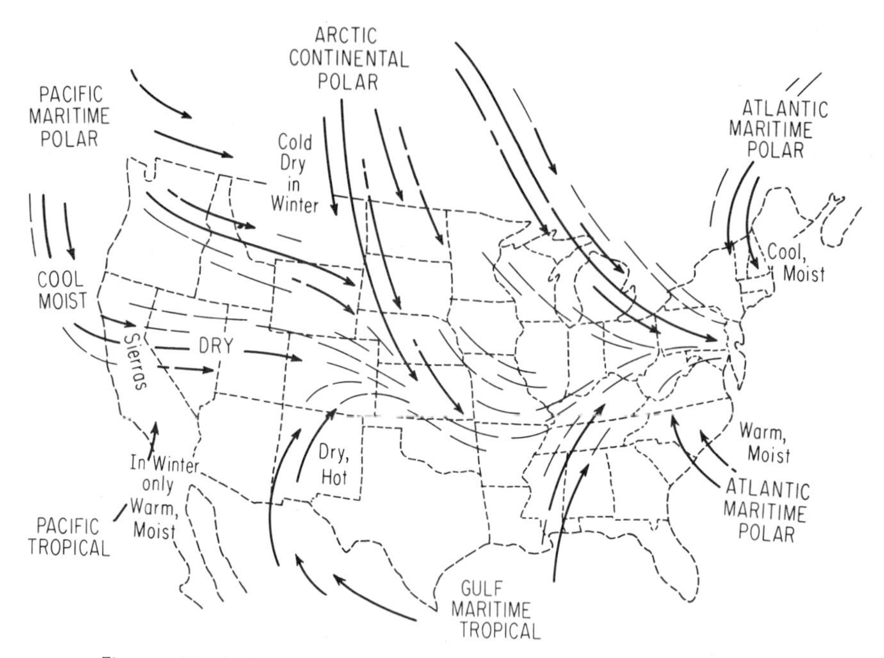

Figure 1.20 Major movements of air masses across the United States.

parts of the west coast or snow in the higher mountains. As such an air mass climbs eastward across western mountain ranges, it loses much of its moisture by precipitation (Fig. 1.21) and eventually reaches the plains states as a relatively dry air mass, sometimes moving on to bring mild, pleasant air to the eastern states.

Of major importance to the weather of central and eastern United States are the maritime tropical air masses that originate in the Gulf of Mexico and adjoining parts of the Atlantic. This warm and moist air tends to move northward up to the Mississippi Valley frequently bringing rain to the nation. Severe weather often occurs where this moist air meets the mild, dryer air that has moved across the country from the Pacific or the cold dry air that has moved down from Canada.

Figure 1.21 Rainfall probability increases as moist Maritime air moves across the Sierras and rises in elevation. The resulting drier air descending along the eastern slopes of the Sierras causes the arid or desert conditions of much of the southwestern states.

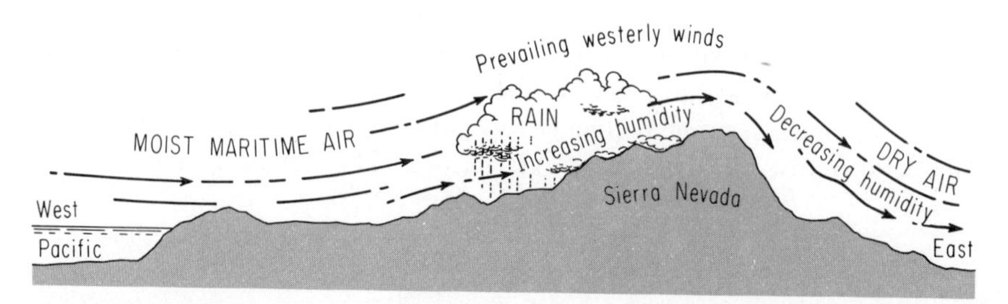

The North Atlantic is the source region for maritime polar air masses that may invade northeastern United States on occasions. This produces the "northeaster" with its drizzling rains or heavy snowfalls but these incursions are rare because of the prevailing westerly winds which usually steer Atlantic air masses toward Europe.

Weather Fronts: Because of their different properties, air masses tend to have distinct boundaries, the well-known *weather fronts.* Along most weather fronts, one mass of air is moving into a region occupied by an unlike, slower air mass. Because of their different densities, the rule of weather fronts is that the air of the colder mass being denser stays close to the ground, causing the warmer air to slide over it and rise upward (Fig. 1.22). Staying close to the ground, the advancing cold air will form

Figure 1.22 Rainfall often occurs along the margin (front) of an advancing cold or warm air mass.

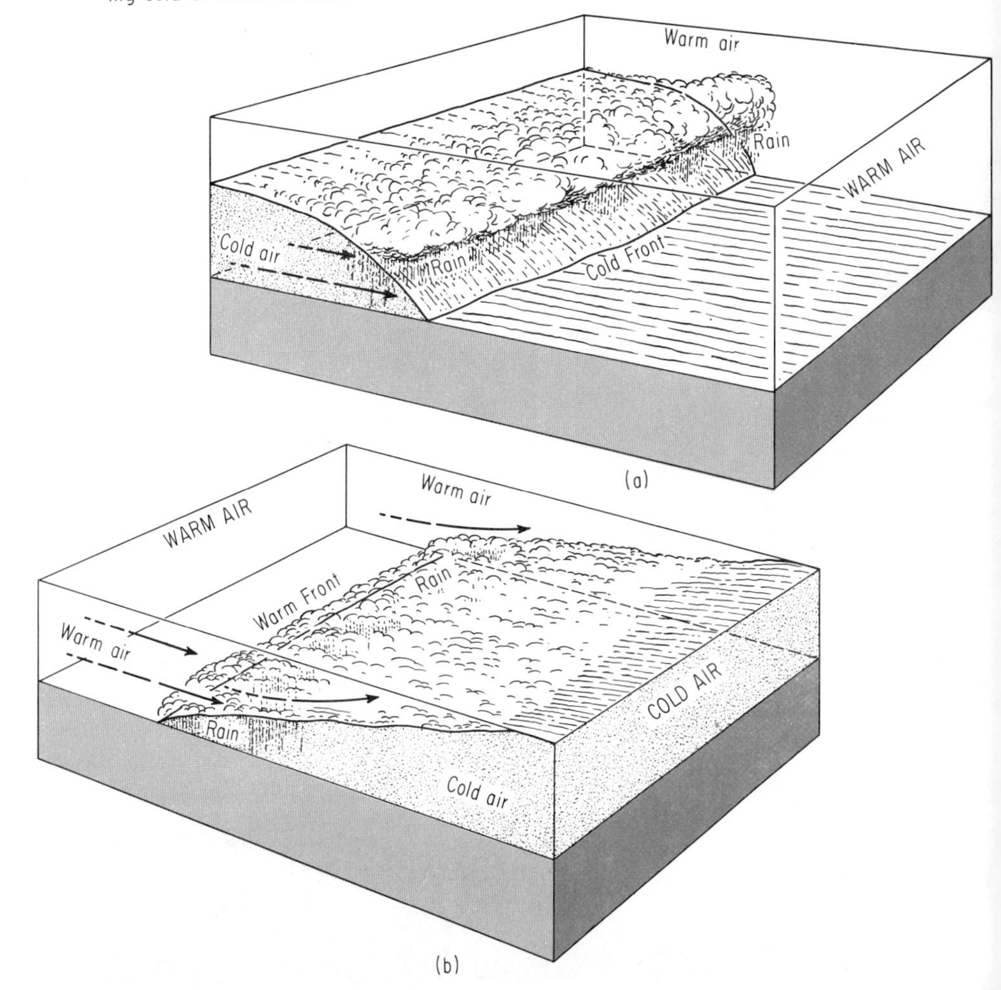

a wedge, pushing the warm air above its advancing edge. If the warm air is of maritime tropical type, it may be unstable enough to break into spontaneous convection units, producing dense cumuliform clouds soaring to extreme heights and thunderstorms. A cold front tends to produce a long line of storm clouds, hundreds of miles long. Not all cold fronts are accompanied by violent storms, but cloudiness is normal and the appearance of the front in a region brings a marked drop in temperature and humidity.

A warm front is formed by a relatively warm air mass moving into a region occupied by colder air. The cold air tends to remain close to the ground, while the warm air slides up above it on a gently-sloping front. Layers of clouds often result from cooling and condensation within the overriding air layer (Fig. 1.22).

The movement of air masses is an essential element in the periodic trapping of air pollutants in many regions. This concept will be developed below.

Altering Weather and Climate

Weather Changes: In addition to changing the chemistry of the atmosphere, modern man has been unintentionally altering the weather in a number of regions, especially in the vicinity of heavy population or industrial concentrations. The emission of particulates from smokestacks around which atmospheric humidity can condense is frequently the cause of fog formation. This is especially common in the Appalachians where factories are often located in narrow valleys surrounded by ridges. A number of wood pulp and paper mills in Washington State regularly produce their own clouds downwind of their stacks.

Cities commonly have heavier cloud cover and record higher rainfall than surrounding regions. This phenomenon is probably related to the effect of the heated urban air raising already humid air to greater heights where it is less able to contain moisture, and the particulate pollutants that serve as nuclei to "seed" the clouds and precipitate the moisture as rain or clouds (Fig. 1.22).

The increase in precipitation in the vicinity of cities is especially

Figure 1.23 Heavier snowfalls east of Lake Erie seem to be caused by the effect of warm particle-laden air produced by lakeside cities over which moist air moves on its way east.

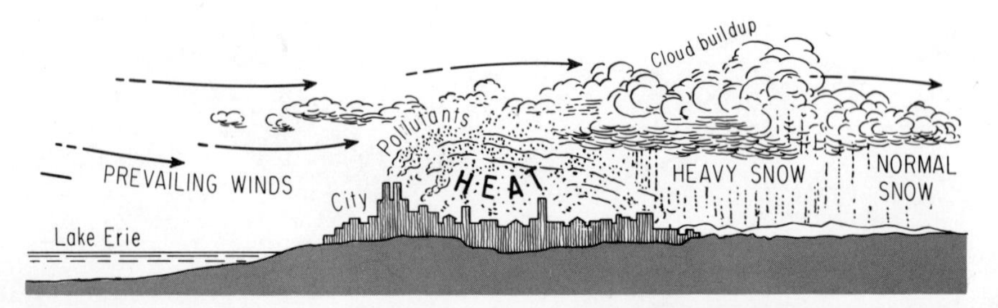

apparent during the winter months. Studies made in New York State and around Lake Erie show heavier than normal snowfalls in the larger cities and in zones extending in the direction of the prevailing winds from these cities (Fig. 1.23). Buffalo, for example, has had a number of paralyzing snowfalls caused by precipitation from humid air moving across Lake Erie and then passing over the city where it is affected by the warm, polluted air.

World Rise in Carbon Dioxide: The clearest and probably the most discussed alteration of the atmosphere by man is the rise in carbon dioxide throughout the world. This is the direct result of the reliance of modern civilization on the combustion of fossil fuels. In the past century there has been a 10 percent increase in atmospheric carbon dioxide and the rate of increase is rising rapidly. About $\frac{1}{3}$ of the amount of the gas produced by combustion has been accumulating in the atmosphere; the remainder is taken up by plant life and absorbed by the oceans.

The rising carbon dioxide levels in the atmosphere present no direct health hazard to man at this time, but there is clearly a potential for long-term climatic alterations. Carbon dioxide is nearly transparent to visible light and has little effect on the incoming solar radiation. However, it is a strong absorber of infrared radiation given off by the land, sea, and clouds. Carbon dioxide then radiates back to the surface some of the absorbed energy (Fig. 1.24). In this way a part of the heat that would have been lost to space remains to warm the lower atmosphere and the surface of the planet. This is sometimes termed the "greenhouse effect" (but a greenhouse also warms by cutting down on wind).

In a sense, the carbon dioxide acts as a partial, one-way filter, permitting the sun's rays to pass through but retaining some of the heat reflected back from the earth. This disruption of the heat balance should have the long-term effect of raising the temperature of the earth, if no other factors interfere.

World Rise in Particulates: Climatological data indicate that from 1880 to 1940 the average world temperature increased by about 0.6° C whereas in the following 30 years the average temperature decreased by about 0.3° C. If we assume that the 60-year warming trend up to 1940 was chiefly the result of combustion causing a carbon dioxide buildup, then there must be other important factors tending to cool the earth's surface; and their effect would have to be strong enough to counteract that of the carbon dioxide buildup of the last 30 years or more during which combustion has been increasing yearly.

One answer may be that urban, industrial, and agricultural air pollu-

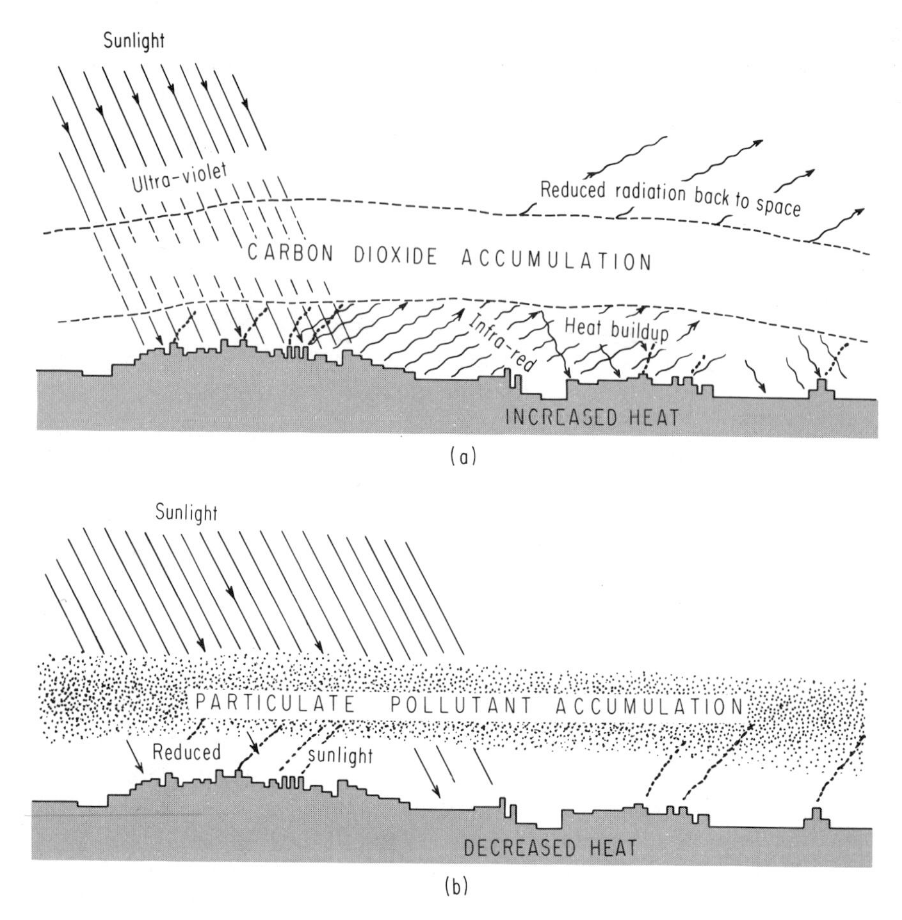

Figure 1.24 (a) Carbon dioxide buildup in the atmosphere may increase temperatures at the surface (greenhouse effect). (b) Pollution buildup in the atmosphere may decrease temperature at the surface by screening solar radiation.

tion includes so much particulate matter as to decrease the transparency of the atmosphere and partly shield the planet from solar heating (Fig. 1.24). Added to this may be the indirect effect of the pollutant particles which may serve as nuclei of precipitation, thereby increasing the amounts of fog and clouds. Such clouds reflect back some fraction of the solar radiation into space. In some areas particulate pollution from jet aircraft has already noticeably increased the high-level cloudiness.

Clearly, the effect of particulate pollution is directly opposite to that of the heat absorption by the buildup of atmospheric carbon dioxide, but we do not know how extensive such heat decrease may become as the particulate pollution of the atmosphere intensifies or by how much the earth's temperature could be lowered.

We should also bear in mind that, although much investigation, and more speculation, is taking place concerning the relative climatic effects of the rising particulate pollution and carbon dioxide levels of the atmosphere, there may still be other factors contributing to the changes in global temperature of the past century. For example, we know little about the significance of the direct heating of the atmosphere, rivers, and oceans (thermal pollution) by power generation (see Chap. 3) and other needs of the everincreasing numbers of humans. Nor can we assess as yet the significance of altering the rate of transfer of heat between air and oceans by growing oil pollution at sea (see Chap. 4).

There may be still other factors, geological processes entirely independent of man, causing long-term changes in atmospheric temperatures more significant than those considered above. The reader is referred to geological textbooks, especially those that deal with modern theories for the Ice Age climatic changes and the shifting of poles in the geologic past.

GEOGRAPHY OF POLLUTION

Clearing the Air

Winds: Air pollution can be detected almost everywhere in the country but is concentrated in the vicinity of large cities or industrial regions. The actual levels of pollutants that can be measured in any one of these emission centers depends on such factors as the population density, the relief of the surrounding region, and weather factors.

Earlier in this chapter we considered the reasons for the large-scale circulation of the atmosphere and traced the movement of air masses across the country. The existence of air masses of contrasting pressure and temperature characteristics is the principle cause of the local winds or breezes vital to the dispersion of smoke and other air pollutants produced by cities (Fig. 1.6). Without these air movements, life in the larger cities would soon become impossible.

It should be kept in mind, however, that air movements or winds sometimes can have negative effects on air pollution levels. Pollutants carried on prevailing winds from a city may contaminate the air of suburban or rural regions miles away. The downward members of twin city groups may suffer greater pollution than its upwind neighbor. Because of prevailing westerlies, New York City's air pollution problem is worsened by the almost constant addition to its air of the products of oil refineries, chemical plants, and other industrial sources at Newark and other towns located just to the west in New Jersey.

Some regions face air pollution problems because of prevailing winds that carry pollutants in from other countries! Southern Sweden, for example, has been receiving sulfurous pollutants carried from the industrialized Ruhr Valley of Germany by southwesterly winds.

A remarkable example of long distant transport by winds was recently discovered when analysis of soils in the West Indies showed a distinctive, reddish-brown silt that originated in the Sahara Desert in Africa. This fine material was carried by the equatorial belt of prevailing easterly winds described earlier in this chapter. Long distance transport of airborne particles was first noted when the ash produced by the great explosion of the volcano Krakatoa in 1883 was carried completely around the world on high level winds. Needless to say, such long distance transport of eroded particles and ash also occurs with man-made particulate pollutants, although the distribution of such materials is poorly understood.

Upward Dispersion and Inversion: Earlier in this chapter we noted the gradual upward drop in temperature that occurs throughout the troposphere. If the ground is sufficiently heated by the sun, the air near the ground expands and tends to rise, bearing pollutants from smokestacks, traffic, and other emission sources. The upward-moving air can then be replaced by cooler and cleaner air from above. This is the normal situation that prevails during the day in most regions (Fig. 1.25). Any breezes serve to increase the effectiveness of this process of vertical mixing and pollutant dispersal.

At night, the ground tends to cool more rapidly than the air and so cools the air near the surface, thereby reversing the usual daytime vertical temperature distribution. The temperature will now increase with height, producing a nighttime low-level "temperature inversion." Instead of rising, the cool, pollutant-laden air may remain near the ground, causing a buildup of pollutants if winds are inadequate to move the air horizontally. Nighttime surface inversions usually dissipate soon after the sun rises and the ground is heated, but if the pollution haze is dense enough it may fail to be dispersed effectively and remain long into the day (Fig. 1.26).

High-level temperature inversions can be a more serious matter. An example is the inversion that develops frequently along the Pacific coast of North America during the warm months. A warm, high pressure air mass usually lies off the coast and tends to move in over western and central California at altitudes of 2000 to 3000 feet, trapping coastal air that has been cooled by the ocean at lower levels. These high inversions may persist for days causing a buildup of pollutants below the layer of warmer air (Fig. 1.27). This is one of the reasons that Los Angeles has had such frequent and severe episodes of air pollution.

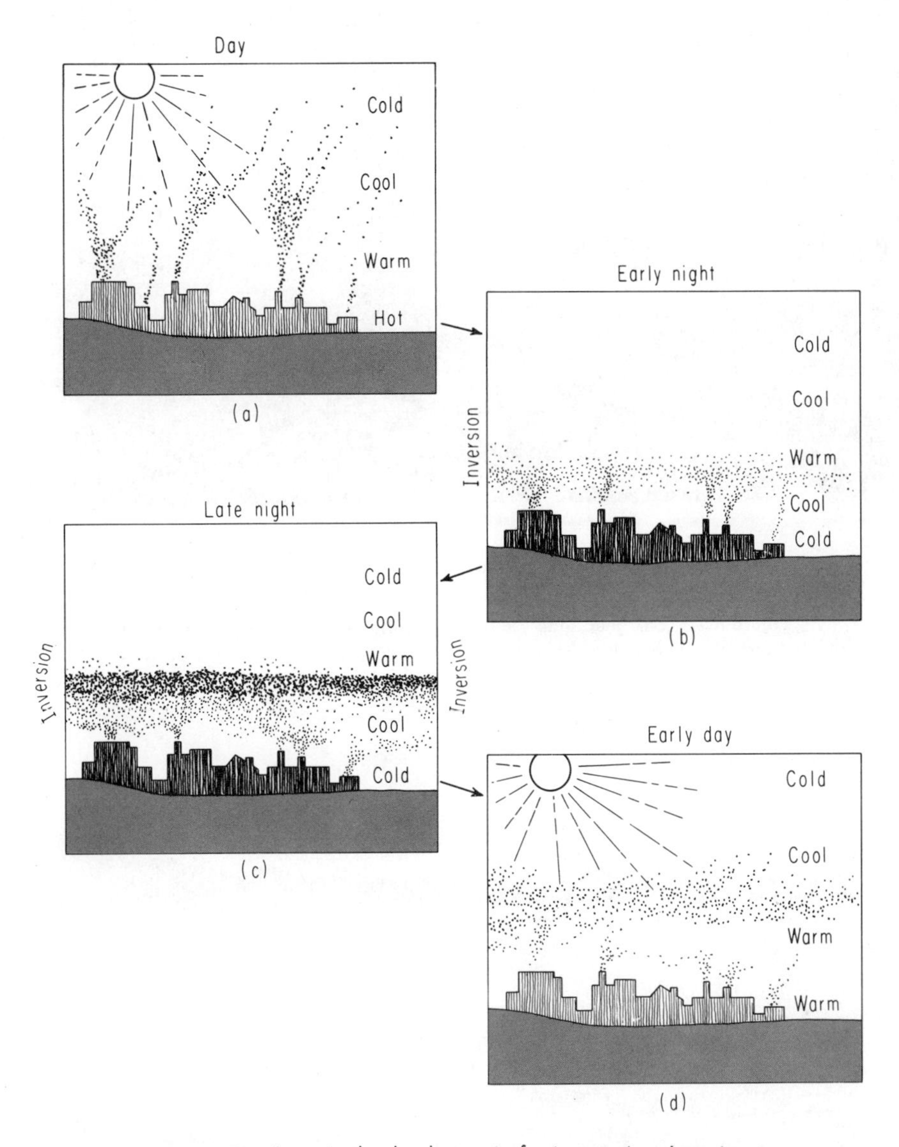

Figure 1.25 Steps in the development of a temperature inversion.

Figure 1.26 Pollutants trapped over Salt Lake City by temperature inversion. (U.S. Department of Interior, Bureau of Land Reclamation.)

Figure 1.27 Smog episode caused by Maritime air moving over Los Angeles and underlying warmer air; the result is an inversion and severe buildup of pollutants.

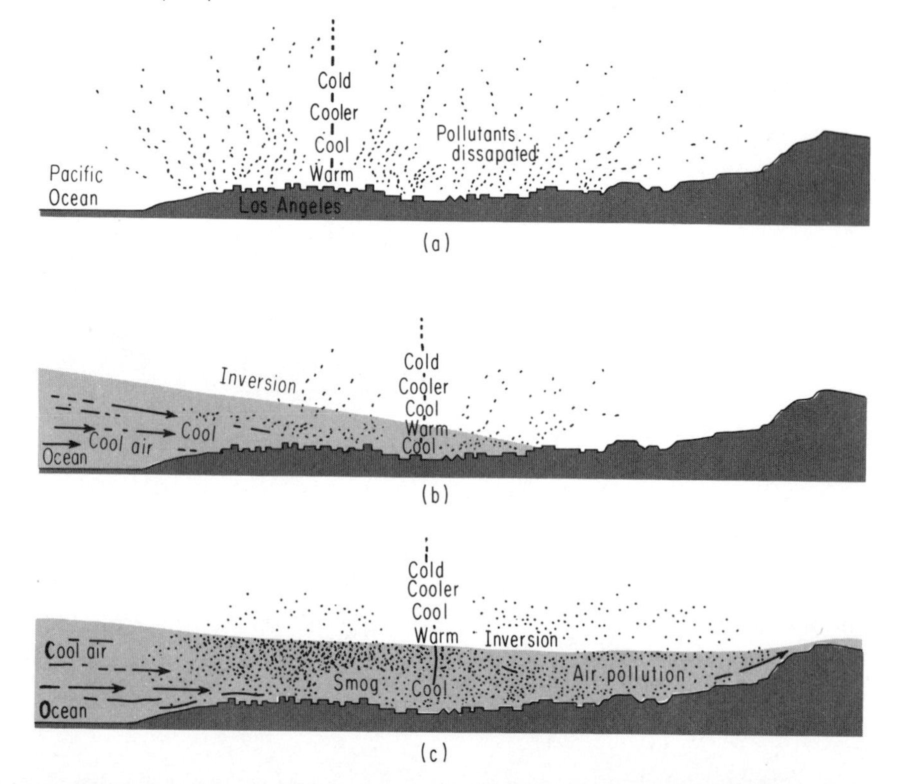

City Air

Most of our population now lives in large cities, subject to varying concentrations of the air pollutants already described. So far, few of the attempts to decrease the level of pollution in cities have met with success. In most cities, pollution levels have risen rapidly in recent years in proportion to the demand for energy and other trends inevitable in our "affluent society."

Attempts to Improve City Air: Los Angeles has clamped down drastically on incineration and power generation within city limits and on other sources of pollutants. Yet the pollution trend is upward because of the inability to effectively control automobile use. New York City has cut the use of sulfurous fuels and has reduced the sulfur dioxide in the atmosphere. Yet its particulate pollution has gone up, including the levels of lead in the air which on many days in 1972 was as much as three or four times greater than the maximum considered safe by the World Health Organization.

London has probably experienced more episodes of fatal air pollution than any other city. The air pollution in London consists of what might be called old-fashioned pollutants, chiefly particulates and sulfur dioxide. These are produced in large quantities by coal burning and have long been typical of London and other large industrial cities. To some extent, fuel oil has replaced coal in London, but the sulfur content of the fuel is still high. In recent years, air pollution episodes have been less acute in London because of regulations concerning fuel, smokestacks, and other factors. These measures have served to decrease the particulate haze enough to allow sunlight to help convection dispersal of fog, thereby minimizing pollution accumulation.

The level of sulfur dioxide in the air of Mexico City has now reached a level exceeding that of London. With very high levels of other pollutants, like carbon monoxide, Mexico City has one of the world's worst air pollution problems.

Smog: The "smog" for which Los Angeles is famous is caused by sunlight acting chemically on automotive waste substances in relatively dry air, as opposed to the natural type of fog experienced in London. Los Angeles has a sunny climate and four million high-powered cars, often traveling at high speed on the freeways. The hydrocarbons and nitrogen oxides given off in quantity by each operating car undergo chemical changes in the sunny climate. Ozone is produced from nitrogen dioxide and some of the nitrogen oxides react with hydrocarbons to form oxidants like PAN (peroxyacetyl nitrate).

The worst pollution danger takes place when a high-level temperature inversion occurs, as previously described. At such times, directives are issued to limit outdoor exercise, particularly by children, which would cause especially heavy intake of polluted air into bronchia and lungs.

A number of American cities other than Los Angeles now experience pollution episodes in which photochemical smog plays a major part, with automobile exhaust providing the basic chemicals. Photochemical smog is spreading overseas as automobiles become ever more abundant. Air quality has been deteriorating especially rapidly in the industrialized parts of Japan, including Tokyo and Yokohama. In Tokyo, the smog (smoggu) has had a high content of sulfuric acid along with photochemical products similar to those of Los Angeles and has been severe enough to strip leaves from the trees. Milan, Italy is another city where smog, produced chiefly from automobile exhausts, has often reached the danger point. Its high rates of chronic emphysema and bronchitis, as well as part of its cancer cases, have been blamed on smog.

Relative Urban Mortality: We have considered the most clearly air pollution-related illness, bronchitis-emphysema and the probable susceptibility to heart diseases of those who experience such respiratory problems. Lung cancer, which has been clearly related to tars and other substances in tobacco smoke, seems to be more likely to develop among those repeatedly subject to various air pollutants, including asbestos fibers and a variety of other particulates believed to be carcinogenic. There can be little doubt that the probability of contracting diseases of the lungs, heart, and other organs rises with the degree of exposure that an individual has to heavily polluted air and would be expected, therefore, to vary with his or her place of residence.

Consider the relative mortality rates between two American cities: relatively unpolluted El Paso, Texas, with a population of about 350,000 and Boston, which has high air pollution levels and a population of about 700,000. The *percentage* of the population of Boston that dies from respiratory system cancer each year is more than twice as great as the percentage of the El Paso population that dies of that same disease each year. For arteriosclerotic heart disease, the ratio is still greater; about *four* times as large a percentage dies from this disease in Boston than in El Paso.

Any Remedies?: The atmosphere is not equally polluted everywhere. On the contrary, the quantity of pollutants ranges from high, mountainous regions where man-made substances are still scarcely detectable to dim city streets where, on certain days, the air is too dangerous to breathe.

An obvious, if impractical solution to localized concentrations of pollutants would be to carefully equalize the spread of the substances, trying to load the air with a specific pollutant at just the right place at the right time. This would be impossible today, considering the present largely-unplanned locations of industrial and power plant smokestacks and the vast numbers of automobiles already concentrated in and around our cities. Even if air pollutants *could* be equalized, we would still face the fact of the growing load of pollutants in the atmosphere as a whole because of rising population and the soaring power demands of affluent living.

Until public awareness and technology can drastically alter the broad pattern of air pollution emission and, perhaps, the basic pollution-creating processes themselves, we must deal with the problems of cleaning up the major sources of present pollution as much as possible.

Cleaning Up the Stacks

Smokestacks used by industrial factories and by fossil fuel-burning power plants account for most of the sulfur oxide and the particulate pollutants produced in the United States. The purpose of stacks is to raise these or other polluting substances high enough so that the pollutants are carried by winds clear of any buildings in the surrounding region. The taller the stack, the more effective is the dispersion. The average height of recently built coal-burning plants is over 600 feet and stacks as high as 1000 feet have already been built. But raising the stacks of an industrial or power plant is only a partial solution to pollution problems. In the total absence of wind, the highest stacks may not prevent the settling of particulants and other pollutants in the vicinity of the facility. And a stack tall enough to disperse pollutants most of the time may be transferring the undesirable substances to a ground location miles away (Fig. 3.26).

Particle Removal: The most effective way of removing suspended particles from stack emissions involves the use of *electrostatic precipitators.* They are widely used in fossil-fuel power plants and many industries. In these devices a negative electrical charge is given to the particles which are then drawn to positively charged surfaces for removal by mechanical means (Fig. 1.28). Over 99 percent of particles passing through the stacks can sometimes be collected by electrostatic precipitators.

In coal-burning power plants the greatest effectiveness of precipitators is achieved with the higher sulfur coals. It is ironic that using lower sulfur coals in order to decrease the sulfur oxide gas pollution tends to decrease the effectiveness of the removal of particle pollutants.

A variety of mechanical collecting devices have also been used to re-

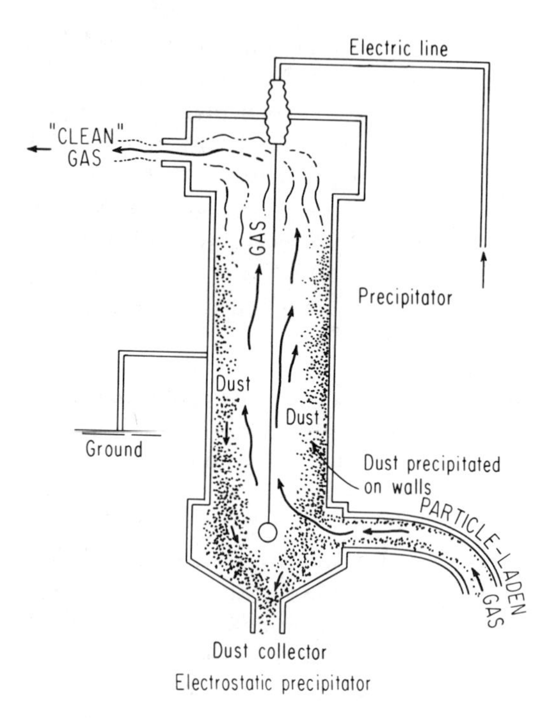

Figure 1.28 Diagram of an electrostatic precipitator used for decreasing stack emissions.

move larger particles from smokestacks. These include different kinds of vacuums, filters, and centrifugal collectors. None are as potentially effective in removing particulate pollutants as electrostatic precipitators.

Gas Removal: Pollutant gases cannot effectively be removed from stack emissions by mechanical or electrostatic devices because of their small, light molecules; their removal requires more complex chemical procedures. For example, sulfur dioxide can be collected by devices using pulverized limestone (or dolomite) which is heated to drive off carbon dioxide, leaving calcium oxide which reacts with the sulfur oxides, thereby forming particles of sulfur compounds which are regularly removed along with fly ash particles, by electrostatic or other means. Some of these devices are called scrubbers.

Nitrogen oxides are more difficult to remove from stack emissions than sulfur oxides. The higher the temperature involved in the facilities, the more nitrogen oxides are formed so that a prime approach has been to try to reduce the peak combustion temperature by a two-stage process of combustion.

Gaseous pollutants can also be removed by absorption by various solids, such as activated carbon and by the use of catalysts that can convert both sulfur and nitrogen oxide into substances more readily removed than either of the two gases.

In summary, we can make the following observations relative to cleaning up stack emissions:

1. Higher stacks can usually reduce local pollutant buildup but do not reduce the pollutants discharged into the atmosphere as a whole.
2. Particulate pollutants can be removed from emissions much more effectively than gases; electrostatic precipitators are most effective.
3. Procedures to remove gaseous pollutants are complex, expensive, and still far from wholly successful.

It should also be noted that it is very common for electrostatic precipitators, collectors, scrubbers, and other pollution-reducing devices installed in power plants and industrial facilities to *decrease* in effectiveness as the operation continues. Often the devices are installed to satisfy local ordinances and are then given little attention or neglected because of the difficulty or expense of maintenance.

Cleaning Up Automobiles

The automobile is one of the chief fascinations of Americans but may be their greatest enemy, considering the 50,000 persons killed each year in collisions and the uncountable thousands endangered by automobile pollution.

The more than 100 million gasoline-powered vehicles on the roads today comprise the greatest source of air pollution in the United States (Fig. 1.8). Legislative control of automotive pollutants began in California with devices to control crankcase emissions which were required on new 1961 model cars. Regulations in the state have become stronger since that time and have served as a guide to the nation as a whole. Nevertheless the total pollution from automobiles in California has been rising faster than effective control devices are applied.

The tail pipes at the back of automobiles constitute the principal source of air pollution in this country. Automotive exhaust systems produce most of the nitrogen oxides, hydrocarbons, carbon monoxide, and lead in our air.

There are two main approaches to reducing the hydrocarbon and carbon monoxide components of exhausts: the exhaust manifold thermal reactor and the catalytic converter. Exhaust manifold reactors allow exhaust gases enough time at high temperature and provide enough oxygen to achieve a higher degree of oxidation, thereby transforming the hydrocarbons into water and carbon dioxide, and converting carbon monoxide to dioxide. The catalytic converter uses a catalyst to promote the more complete oxidation of hydrocarbons and carbon monoxide.

The problem in using the manifold thermal reactor is the higher temperatures achieved increase the production of oxides of nitrogen so that using this device reduces carbon monoxide and hydrocarbons but increases the nitrogen oxides. Use of the catalytic converters means dealing with several problems. Their operation requires a fairly constant temperature that is usually not achieved until the engine has been running for a long time; also the catalyst surface tends to become coated with the lead which is part of most gasolines used today, thereby reducing the effectiveness of the converter.

Methods of reducing emission of nitrogen oxides in automobile exhausts were still in the research stage in 1973. Lower temperatures mean less nitrogen oxides, and one approach has been to recycle part of the exhaust gas back into the engine in order to reduce the peak combustion temperature at which most of the nitrogen oxides will be produced.

The elimination of lead compounds from automobile exhausts seems to be impractical by any means other than by totally ending the use of lead in the gasoline itself. There has been great resistance to this approach because of the usefulness of lead compounds in improving the octane of a gasoline at low cost.

Most of the exhaust control systems now being used or proposed for use in this country are of the manifold thermal reactor types, sometimes called "afterburners." Maintained properly, they can nearly eliminate emissions of hydrocarbons and carbon monoxide. Unfortunately, there may be an *increase* in nitrogen oxides and the lead emissions are essentially unchanged. If unleaded gasoline is used in an automobile with a properly adjusted and well-maintained emission–control device of this type, only nitrogen oxides may remain as the almost unavoidable byproducts of internal combustion.

Cleaning Up the City Air

We have already examined the reasons why our cities have major air pollution problems. The pollution-producing conditions are so complex and have developed so gradually that many environmentalists consider the problem insoluble except by the most drastic measures.

Since automobiles are the worst polluters of all, many European cities have simply closed a number of streets to car traffic. Swedish officials have been considering banning private automobiles entirely from the downtown areas of cities of over 100,000 persons. It seems likely that New

York City will eventually have to close down the inner districts to automobile traffic because of the high levels of carbon monoxide, hydrocarbons, lead compounds, and other pollutants on the clogged city streets.

The heavier the traffic in a city, the slower the cars must go and the higher the amount of carbon monoxide and hydrocarbons produced by the engine. This is an argument for restricting traffic to a few large roadways if it must pass through the city at all. If private cars are to be banned from the center of a city, there must be effective rapid-transit systems available to the public. Unfortunately, many cities long ago abandoned their rapid-transit railways, trolley lines, and electric buses, leaving inner city transportation to gasoline-powered buses, taxis, and private cars, polluters all.

There is no doubt that air pollution in city centers can be greatly reduced by limiting or ending private car traffic. Alternative proposals include installing giant fans to suck air pollutants out of the streets, possibly through existing sewer pipes, restricting private transportation to electric vehicles in the inner city, and the development of a network of moving sidewalks.

The automobile has been the target of a number of state laws and numerous proposals in the United States Congress. Recent years have seen a fencing match between the federal administration and the automobile manufacturers, the result being agreements to install exhaust-control devices of increasing efficiency on new model cars. Emission controls on new model automobiles are now a reality and will undoubtedly become more efficient in coming years. But there is great doubt that these will be maintained adequately.

In 1973 well over 90 percent of American automobiles burned leaded gasoline, passing most of this lead out into the nation's air; most cars had only minimal pollution control devices that were usually operating well below top efficiency, serving to decrease but not eliminate hydrocarbons and carbon monoxide, and almost all of our cars produced uncontrolled amounts of nitrogen oxides.

Suggested Readings

American Chemical Society, *Cleaning Our Environment: The Chemical Basis for Action,* A report of the A.C.S., 1969.

BRODINE, VIRGINIA, "Point of Damage," *Environment,* Vol. 14, No. 4, 1972.

DARLING, F. F. (ed.), *Implications of the Rising Carbon Dioxide Content of the Atmosphere,* Conservation Foundation, Washington, 1963.

FLOHN, HERMANN, *Climate and Weather,* McGraw-Hill, New York, 1969.

FROST, JUSTIN, "Earth, Air, Water," *Environment,* Vol. II, No. 9, 1969.

HULSTRUNK, A., *Air Pollution,* Associated Press, New York, 1969.

LEWIS, HOWARD R., *With Every Breath You Take,* Crown Publishers, New York, 1965.

MARTIN, A. E., "Mortality and Morbidity Statistics and Air Pollution," Symposium No. 6, *Proceedings of the Royal Society of Medicine,* **57**, 1964.

MITCHELL, J., *Global Effects of Environmental Pollution* (S. F. Singer, ed.), AAAS Symposium, 1969.

RIEHL, H., *Introduction to the Atmosphere,* McGraw-Hill, New York, 1965.

STRAHLER, ARTHUR, *The Earth Sciences,* 2nd ed., Harper and Row, New York, 1971.

VAN DUUREN, BENJAMIN, "Is cancer airborne?," *Environment,* Vol. 8, No. 1, 1966.

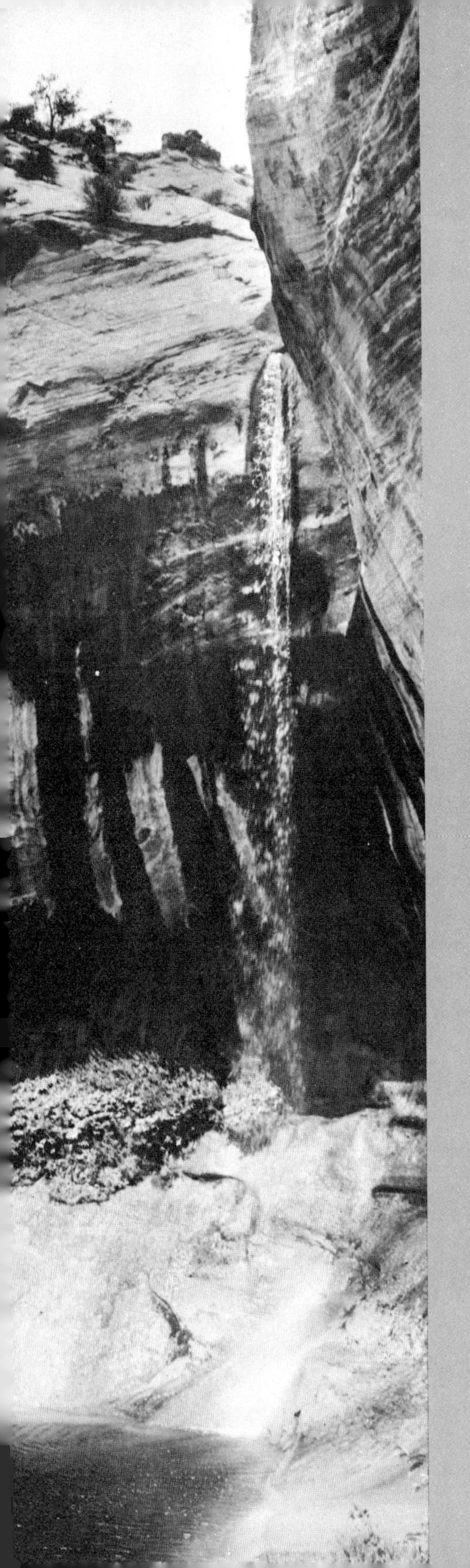

2 Water Supply

Union Pacific Railroad Photo.

To sustain a human life in a physiological sense, about one gallon of water a day is needed. But nowadays in the average American city the usage totals about 200 gallons a day per person. Climate, degree of affluence, and local customs all affect water consumption. In some arid parts of the Southwestern United States' Indian reservations, as in some dry and undeveloped countries, the average domestic usage is less than five gallons a day per person. By contrast, in Beverly Hills, California, with its lawns and swimming pools, the usage is more than 500 gallons a day per person.

The demand for water is increasing almost everywhere in the world for two principal reasons: burgeoning population growth, and increasing affluence which requires more water per capita. The total water usage in the United States has increased more than seven times since 1900.

Consider the change in water use habits in our country. In 1900 a full bath was a luxury, perhaps a once-a-week occurrence, for most people. Today, baths and showers are daily habits; each tub bath requires 25 gallons of water. The outhouse has been replaced by the flush toilet, which requires three gallons of water at each flushing. A dishwashing machine uses twice the three gallons of water needed to wash dishes by hand. Sprinkling a 50 by 100 foot lawn requires 1000 gallons of water. And then there is the waste from leakage: each day in the United States billions of gallons of water are wasted just from dripping faucets. When we add growing industrial needs to growing domestic needs, the demand

Figure 2.1 Irrigation ditches fed by a canal leading from a distant river provide moisture for these crops in dry eastern Washington. (USDA Forest Service.)

for water, especially in the affluent mechanized United States, becomes alarming.

But the ever-rising use of water for agriculture may represent the most serious water demand of all. Unlike the water used by industry (which reenters streams or rivers for possible reuse), about 60 percent of the water used during irrigation evaporates or is otherwise lost (Fig. 2.1). As population constantly expands, so does irrigation to augment crop yields, especially in arid areas. In many regions, irrigation already consumes more fresh water than all other uses combined.

Whether we consider domestic, industrial, or agricultural uses of water, one fact is clear: the demand for fresh water has been increasing every year but the supply has not. In some municipalities in the United States the water supply has dropped to such a critical level that drastic legislative measures are being initiated to eliminate waste, reduce pollution, and conserve the supply. Everywhere, the water situation worsens as population grows and per capita consumption increases.

OUR FRESH WATER SUPPLY

The Renewable Resource

Water has the unusual property of being an essentially inexhaustible natural resource. It is made renewable over and over again by the *hydrologic cycle,* the endless migration of water particles which has been in process as long as there have been oceans and atmosphere. In its simplest form, the hydrologic cycle consists simply of evaporation of sea water,

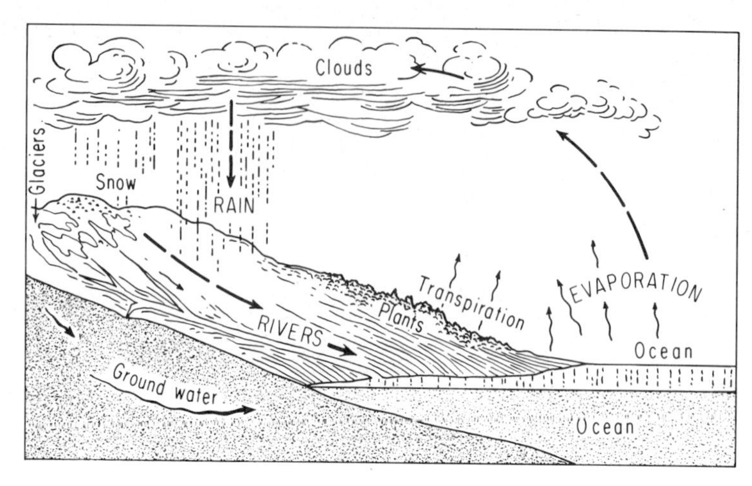

Figure 2.2 The hydrologic cycle. Water evaporates from oceans (or other wet surfaces), vapors in the atmosphere, condenses as clouds that provide source for rain which flows (or snow which melts) via rivers to the oceans, there to evaporate again and continue the cycle.

Figure 2.3 Rain falling from these distant clouds in Texas is part of the endless hydrologic cycle powered by gravity and evaporation. (Photo by author.)

Figure 2.4 Rivers are also a vital link in the hydrologic cycle; water from rains or melting snow is pulled downward by gravity, eventually to reach the oceans, the principal sites of evaporation. Yellowstone River, Wyoming. (Photo courtesy of Union Pacific Railroad.)

condensation as clouds, and precipitation (rain or snow) returning to the salty oceans (see Figs. 2.2, 2.3, 2.4). In effect, this amounts to a vast natural desalinization process. Evaporation also occurs on land from streams, lakes, and other wet surfaces. In all cases the activating energy behind the process of evaporation is the heat of the sun, and the return of water as precipitation is caused by the pull of gravity.

Most rain which reaches the ground evaporates immediately or, after a short-lived sojourn in the upper soil, is absorbed by plants which later give off much of it during their metabolism. Some rain soaks down through the soil into pores and fractures in the bedrock to become *ground water*. Only a fraction of the rain becomes *surface runoff*,

Figure 2.5 When snow falls on high-altitude regions like Mount Rainer shown here, it may become the source of glaciers, which eventually melt to feed their waters into seaward-flowing rivers. Snow and glaciers may be considered delayed segments of the hydrologic cycle. (Photo courtesy of Union Pacific Railroad.)

flowing over the land's surface, forming brooks and rivers, and usually reaching the sea. In cold areas some of the precipitation that falls as snow is stored for some years in the form of glaciers—masses of ice that slowly move to lower altitudes where they melt, thereby continuing the hydrologic cycle (Fig. 2.5).

Water is said to be reusable because it continually becomes available again after use, thanks to the hydrologic cycle. But this remarkable characteristic does not alter geographic and climatic facts of distribution. Even with the continuous renewal process, there is only a finite supply of fresh water during a specific time at any specific location, which places a limit on the number of people who can live there. Sufficient quantities of water are not available everywhere to satisfy recent growth patterns. Nor is the available water always safe enough to use: increasingly, pollution is becoming a serious limitation. In the following pages we will consider sources and limitations of supply, possible ways to increase that supply, and the growing threat of pollution.

Sources of Water Supply

About 97 percent of the earth's water exists in the form of salt water. Only three percent is fresh water, and most of that appears as ice in the

Antarctic and Greenland icecaps and in other glaciers. Considerably less than one percent of the total water in the world is available as fresh water in a form which can be readily used to fill the needs of life and civilization.

Almost all of man's freshwater supply comes from two sources: surface runoff waters and ground water and they are both derived from (and limited by) precipitation from the atmosphere (Fig. 2.3).

Surface runoff water is available in the form of rivers and smaller streams, lakes, and reservoirs.

Ground Water Supply: Ground water from precipitation soaks down through the soil and porous bedrock until it reaches a depth where spaces or interconnections are closed by pressure. Above this limit of penetration lies a saturated zone in which water fills all soil and rock openings. Its "top" is called the *water table* (see Fig. 2.6). To obtain ground water we dig or drill down through the water table into the zone of saturation.

The water table is neither level nor is located everywhere at an equal depth below the surface of the ground, but it is usually deeper below hilltops and shallower on valley sides. Ground water slowly sinks downward, spreads outward, and seeps into the valleys as springs or streams. Heavy rains renew the ground water and thus raise the water table; during droughts the water table sinks because ground water from the saturated zone slowly but continuously seeps out without being renewed. Heavy pumping of ground water from wells sunk into the saturation zone has the same effect as a drought: it lowers the water table.

A particularly effective water-bearing body below the surface of the ground is called an *aquifer.* An aquifer usually is a layer of porous rock adjacent to, or sandwiched in between, more impervious rocks. A better supply of water may be obtained by drilling down to this porous layer than merely to the rock and soil above it. Many vital aquifers are tilted, beginning at or near the surface of mountainous regions and inclining downward to spread out below other regions at lower elevations (see Fig. 2.7). Where wells tap such tilted aquifers, the water may flow upward without necessity of pumping because of the pressure caused by the weight of water at the higher elevation in the distant parts of the aquifer. Wells of this type are called *artesian.* They are of great importance for irrigation in arid regions. An aquifer is replenished by rainfall in its *recharge area,* the region where the aquifer is at higher elevation, and at or near the surface. A decrease of rainfall in the recharge area or excessive pumping both deplete an aquifer.

Surface Water Supply: One of the problems of water supply from surface sources lies in the wide seasonal variation of most rivers' discharge

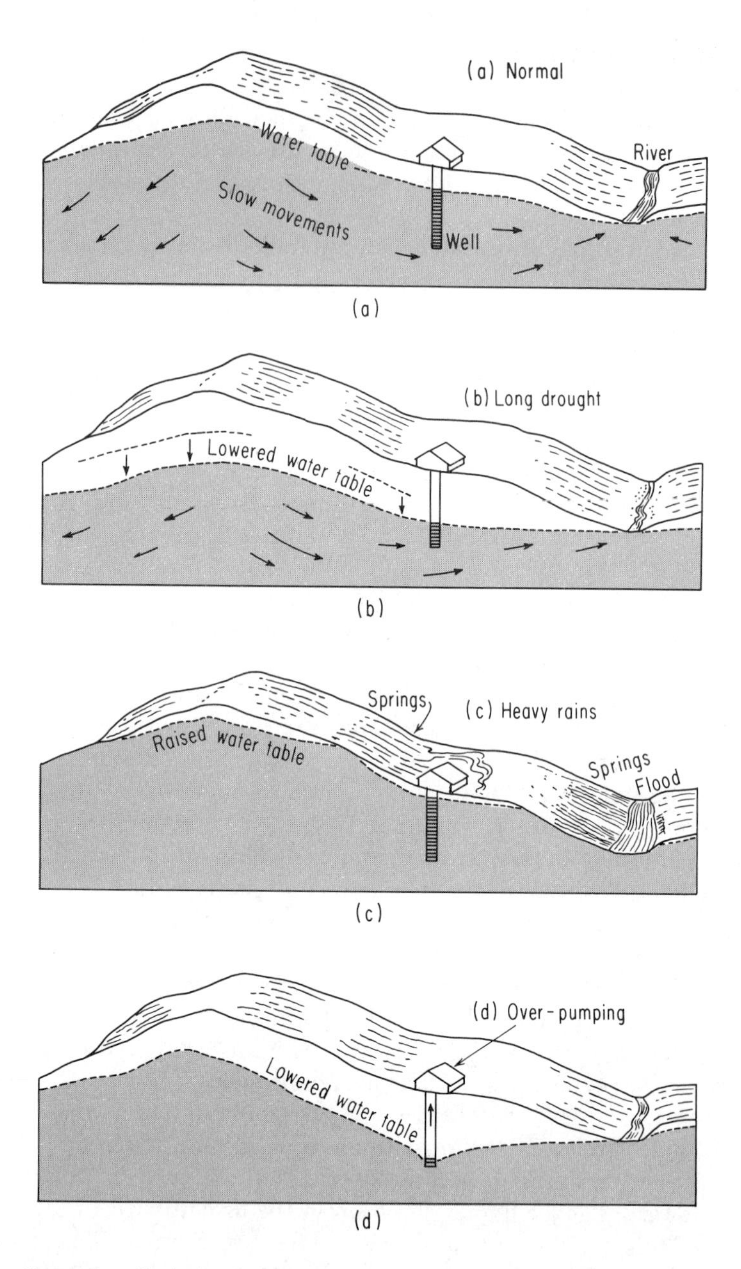

Figure 2.6 Block diagrams illustrating the water table. (a) Water table is the upper surface of the cone saturated by underground water that has seeped down during rainfalls. (b) After a long drought, part of the underground water has flowed out to the nearest river, and the water table is lowered. The river has decreased in flow. (c) Heavy rains add to underground water and raise the water table; springs occur where the table intersects ground surface. (d) Constant or heavy pumping from wells will draw water from below ground, cause a lowering of the water table around the well (cone of depression), and may exhaust underground water supply.

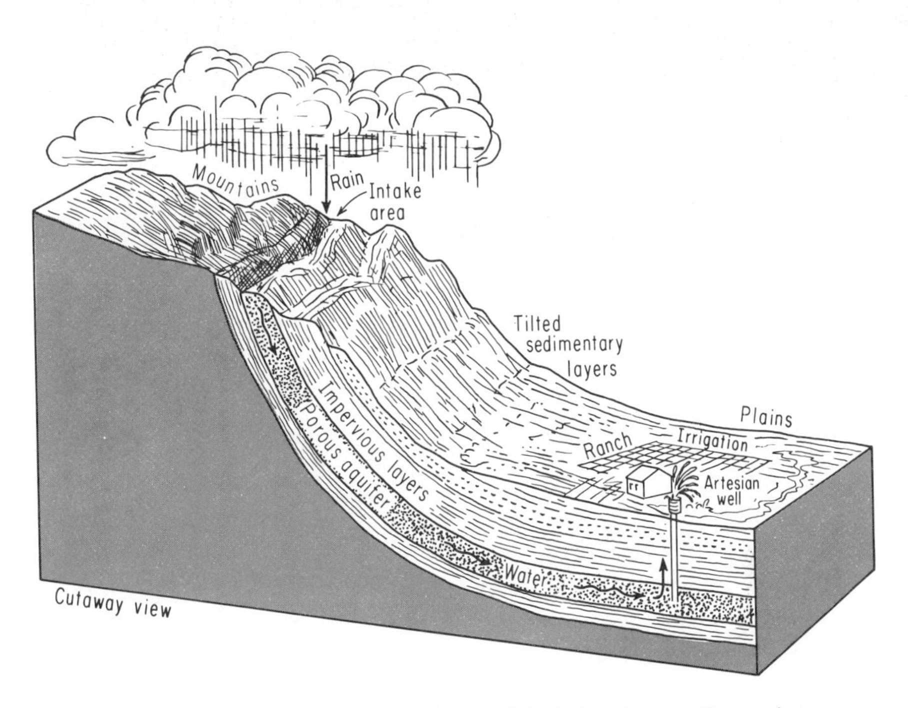

Figure 2.7 Aquifers are porous layers of buried rock or sediment that become saturated with water from rainfall in high intake areas and can be tapped by deep wells to provide artisian water for irrigation value in some areas.

Figure 2.8 A dependable supply of water, both surface and underground, may be threatened by the destruction of forests, and root systems, and humus that tend to control rapid surface runoff of rainwater and regulate the downward seepage of water to the water table. (USDA Forest Service.)

Figure 2.9 The lake formed by Mano Dam, California became filled with silt within a few years after its construction. Such a rate of sedimentation is unusually rapid, but all reservoirs or lakes formed by dams are threatened by eventual filling in by silt carried from upstream, resulting in ultimate loss of their potential for irrigation, water supply, hydroelectricity, and flood control. See Figure 3.7. (USDA Forest Service.)

(the quantity of water passing a given point in a stated length of time) as compared to man's water needs, which remain relatively constant.

Forest cover is one of nature's most important means of regulating water flow through a region. Protective vegetation, especially trees, slows the rate at which runoff occurs and aids its infiltration underground, thereby making it available for wells. But forest cover is being destroyed everywhere (Fig. 2.8). To compensate, engineers build dams to create reservoirs by controlling the rate of runoff and holding back river waters. But reservoirs must be located where engineering considerations and relief of the land allow rather than at major areas of public consumption. The great expense of their construction, as well as the expense of construction of aqueducts to carry the water to population concentrations, plus the speed with which reservoirs fill with sediment, limits their usefulness for water retention and supply (Fig. 2.9).

Another drawback of reservoirs is the great amount of water lost by evaporation from their surfaces, which is considerably more than occurs during the natural flow of the undammed rivers. In an arid climate more than $\frac{1}{5}$ of the water impounded behind a dam may be lost by evaporation. Where the reservoir is used to feed irrigation canals in agricultural regions, the percent of runoff lost to evaporation is even greater. As much as half of the water flowing through irrigation canals may evaporate. We will consider the role of dams for energy production in Chapter 3.

THE SHRINKING SUPPLY

Water Crises Around the World

The volume of available fresh water is ultimately limited by the amount of rain or snow, which is a relatively fixed quantity, dependent on climatic factors. A drastically increased population can accomplish the same thing as a drought: a serious water shortage.

Scientists suspect water shortage as the reason for abandonment of many ancient Mediterranean and Middle Eastern cities where overgrazing and excessive tree-cutting had reduced plant cover and its water retaining capacity, and population growth and heavy well use had lowered the water tables. Fatehpur-Sikri, an architecturally unique city near Agra, India, was abandoned about 1600, partly because water supplies had been exhausted by population growths and droughts.

The cliff dwellers of Arizona and Colorado were forced to migrate after a devastating drought continued for almost the entire last quarter of the 13th century (Fig. 2.10). As the water table slowly dropped, the small springs and spring-fed streams near the foot of the Indians' cliffs

Figure 2.10 The cliff-dwelling Indians of Arizona abandoned their homes over 500 years ago when a drought continued for many years. Montezuma Castle National Monument. (Photo by author.)

dried up, one after another. Most of the cliff dwellers finally departed for the Rio Grande Valley and established communities more difficult to defend but with more dependable water supplies.

In the Old World, the growth of cities with large populations led to the development of aqueduct systems which tapped distant water bodies. The first major example was built more than 2000 years ago to supply the city of Rome. The seven Roman aqueducts, one of which was more than 50 miles long, combined stone-lined surface channels of gentle Rome-ward slope with great arched bridges which carried the water over low regions and valleys without disrupting the gradient. In later aqueduct systems it became possible to eliminate elevated bridges by means of a siphon system; gravity pressure pushed the water, after flowing through tunnels under valleys, up again on the other side. All that was needed was a water source at a higher elevation than the city being supplied. Mechanical force pumps were the next step in making transportation of water more efficient, even where the source was not appreciably higher than the destination.

Many centers of civilization, especially in arid regions, have been dependent on concentrations of good wells just as other cities have been dependent on rivers. Even in nonarid regions as rivers became polluted people at some distance from the rivers turned to wells as a source of water. Nowadays in many communities an increasing use of electric well pumps (especially by industry) has withdrawn so much water from the saturated zone that the water table has gradually dropped, making ever deeper wells necessary (Fig. 2.6).

Rising agricultural and industrial demands for water have caused a serious drop in Israel's water table, especially in the south. But the problem has been temporarily solved by ingeniously pumping water from Lake Tiberias in the north down into wells that reach down into the porous strata underlying much of the country and by increasing the diameter of surface water supply redistribution pipes. Underground water now spreads southward from the Tiberias region raising the level of water in wells in the Negev and other dry southern regions. However, desalinization plants under development may promise the most satisfactory long-term solution to Israel's growing water needs.

Water Crises in the United States Today

Almost everywhere in the United States habits of water usage are based on the assumption that water is free and abundant, if not downright inexhaustible. Most Americans use water without restriction or thought.

United States cities with a population of less than 100,000 usually depend for their water supply on ground water brought up by a system of deep municipal wells. In recent decades most small cities have experienced water supply problems as growing populations and increased per capita water use have resulted in falling water tables. Some cities have increased their supply by acquiring surface systems, impounding nearby or distant streams (Fig. 2.11). Often more than one city claims the same supply of surface water, and conflicts develop.

In cities still dependent on ground water, the supply problem is worsened by the paving of broad areas, which prevents rainwater's seeping into the ground. A total of 100 million acres of ground are paved over annually in the United States; 400 acres were being paved over in California every day during 1971.

At the United States' present rate of population, affluence, and industrial growth, meeting the nation's water needs in the year 2000 could mean using 75 percent of all the surface water in the country. All of the largest rivers would have to be channeled into pipes and aqueducts and our largest lakes would shrink to tiny stagnant ponds.

The shrinking freshwater supply has been most apparent in large cities and in communities of the Southwest with sunny, dry, attractive climates, into which there has been a recent shift of population and industry. Let us consider a few sections of the United States already encountering water supply crises.

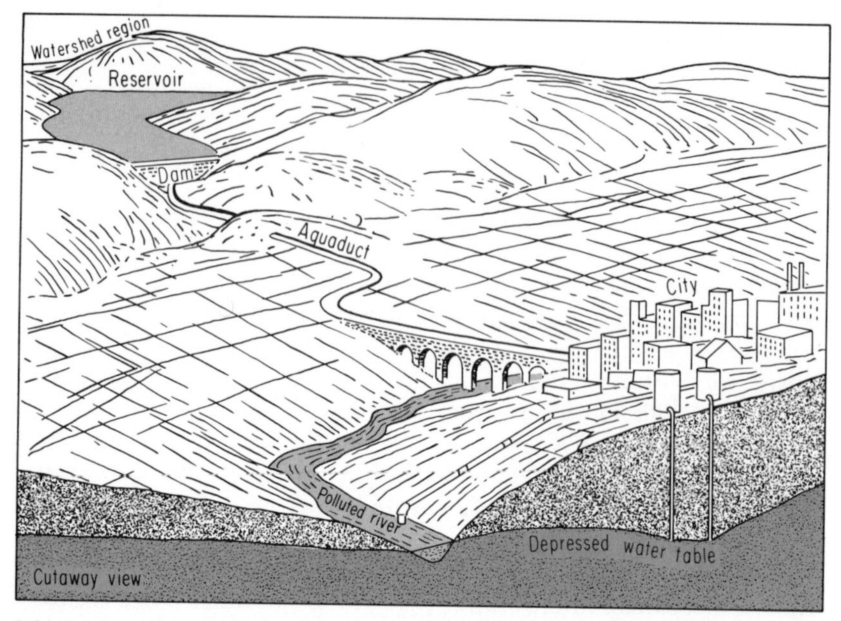

(a)

(b)

Figure 2.11 (a) The city figured here has essentially exhausted its underground water, damaged nearby river water by waste pollution, and now depends on a distant reservoir connected by an aqueduct. (b) The Croton Aqueduct passing over the East River in New York City, part of the original aqueduct-reservoir water supply system of the city. (Photo by author.)

West Texas–New Mexico: West Texas and New Mexico are normally arid. Most of the Rio Grande has a high sediment content; the parallel Pecos River has a high mineral content. The flow of neither one is sufficient for soaring population demands. But irrigation farming, mostly with ground water, has been increasing since the 1940s. In New Mexico large amounts of ground water are blackish and unusable. In many areas ground-water supplies have been so depleted by irrigation, municipal and industrial demands, that there seems little hope of maintaining the present economy (let alone inevitable growth) without importing water soon. Even imported water may not be able to sustain farming in all of the areas now under cultivation.

California: The dry southern part of California with its rapidly increasing population and heavily irrigated agricultural land is facing major water supply problems; the mountainous northern parts of the state enjoy a water surplus. In the south the use of ground water is very heavy, sometimes exceeding the rate of replenishment by natural means. So much ground water has been withdrawn in the San Joaquin Valley that some wells are beginning to draw saline water from the ocean to the west, and land subsidence is occurring in places. An elaborate long-range plan to transfer California water from the wet north to the dry south is in an advanced stage of development, but tremendous expense and ecological dangers are involved. Other alternatives under consideration are: large-scale south coast desalinization plants, the reuse and recharging of ground water, and procedures to be discussed later in this chapter.

Los Angeles: By 1900, the city of Los Angeles was reaching the end of a 100-year dependence on water from the Los Angeles River. In 1914 an aqueduct was opened at great expense to tap the waters of the Owens Valley east of the Sierra Nevada ranges, more than 200 miles away. By the 1920s population had grown so much that the aqueduct system had to be extended north of the Owens Valley to siphon water from Mono Lake. Then in the 1930s Los Angeles reached out to tap the distant Colorado River, by means of an aqueduct-tunnel-reservoir system of considerable complexity. Despite this web of surface water supply systems, well pumping in the Los Angeles area has increased at such a rate that the water table has dropped in places to more than 75 feet below sea level, allowing salt water from the Pacific Ocean to seep as far as two miles inland.

New York City: New York City stands on the banks of the mile-wide Hudson River, which is essentially untapped and useless partly because

Figure 2.12 Industrial waste pollutes the waters of the Hudson River on its way to the Atlantic Ocean. (USDA Forest Service.)

Figure 2.13 The waste from the city of Nashville, Tennessee pollutes the Cumberland River. (USDA Soil Conservation Service.)

of pollution from upstream communities and the city itself (Fig. 2.12). Pollution regulations are weak or unenforced. For its water supply New York City depends on distant reservoirs and a complex aqueduct system which often is inadequate. Yet millions of gallons of water flow wasted from faucets uncontrolled by any domestic water metering system.

Historically, New York City's long dependence on local streams and well water ended in 1842 with the completion of the Croton Reservoir about 25 miles north of the city and a long aqueduct (Fig. 2.11b). The reservoir was enlarged in 1890 by a higher dam, and the water flow increased four-fold by a larger aqueduct. By the 1900s it was apparent that the entire Croton watershed region was inadequate for the burgeoning population. By 1915 several new reservoirs were completed in the Catskill Mountains about 90 miles north of New York City; the mountain water was carried under the Hudson River and to New York City by an aqueduct which was a major engineering achievement. Soon, more reservoirs were added to the Catskill system. In 1945 an additional major watershed area was captured by damming two tributaries of the Delaware River (far west of the city) and constructing the Delaware Aqueduct. Yet even this vast Croton-Catskill-Delaware system has often proved insufficient.

In 1965 one of the East's most serious droughts took place. The city's reservoirs shrank to a level so dangerously low that extraordinary measures were invoked to cut down on water consumption. New York became involved in a conflict with Philadelphia about the amount of water it was withdrawing from the Delaware River, which also serves as Philadelphia's water supply. Before the drought finally ended some 40 million people in the eastern United States had been affected by a severe water shortage. The probability of future shortages is a strong one; New York City cannot keep extending its aqueducts to sources progressively farther away without additional jurisdictional conflicts and tremendous expense.

WATER POLLUTION

The pollution of surface and ground waters is increasing more rapidly than our solutions to the water crisis. About $\frac{1}{3}$ of the wells in the United States no longer produce pure water. One can drink from very few of the rivers. The continual growth of cities together with the rapid expansion of industry is overwhelming the ability of most rivers to carry off wastes. High costs of garbage and water purification treatment are expenditures postponed or avoided by municipal governments, while U. S. waterways become more and more like open sewers. Most river water flowing into the oceans is badly polluted. The damage of such drainage to ocean waters will be considered in Chapter 4.

What Is Water Pollution?

Water is almost never pure in the chemical sense of an unvarying compound of hydrogen plus oxygen. Natural water readily absorbs small amounts of gases, various substances, and a community of living organisms in solution. Water is considered polluted if it is not suitable for intended use (recreation, agricultural irrigation, industrial needs, or human consumption) because of undesirable substances: sediment, organic matter, or chemical substances, now including radioactive elements (Fig. 2.14).

Organic matter discharged into rivers is broken down by bacteria and other microorganisms that require for growth large amounts of oxygen, most of which has been absorbed from the air. Some degree of a water body's pollution may be established by measuring the biochemical oxygen demand (BOD), the amount of oxygen a unit quantity of water absorbs in five days at a temperature of 64° F. When the concentration of polluting substances reaches a high enough level, their breakdown uses up all oxygen in the water. Fish and other animals die of asphyxiation (Fig. 2.15). The oxygen-using (aerobic) bacteria disappear; anaerobic bacteria replace the aerobic forms and may break down sulfate and nitrate compounds to produce toxic chemicals. Even pieces of wood or other undecayed natural matter in large quantities are detrimental to water quality because the oxygen requirement of such matter can be high enough to cause water to lose its purifying potential for other, more offensive wastes.

An indicator of potentially dangerous water pollution is the coliform bacteria content. These bacteria originate in the intestinal tract of warm-

Figure 2.14 Rivers may be damaged by chemical, organic, and thermal pollutants from a variety of sources: domestic, industrial, and agricultural.

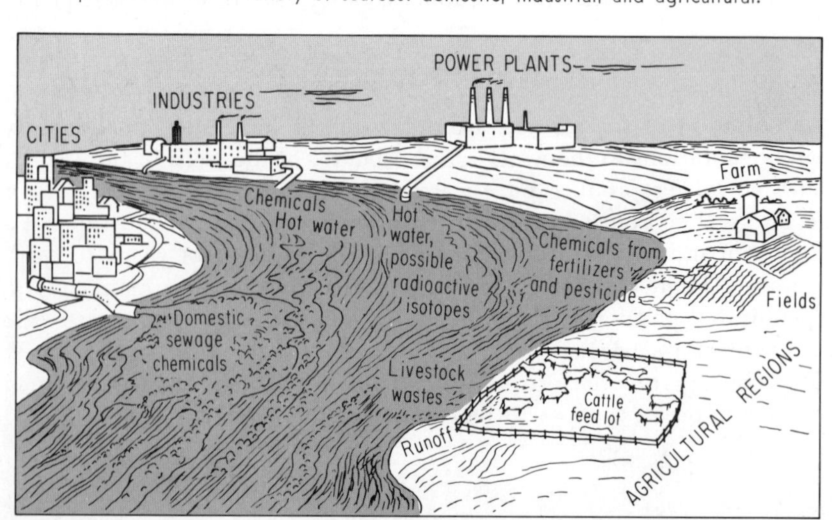

Figure 2.15 When the organic pollutants in a river become sufficiently concentrated, BOD rises to the point of oxygen exhaustion, and life is destroyed. Dead fish near Colorado Springs. (USDA Soil Conservation Service.)

blooded mammals and are always present in new sewage, which may contain typhoid, hepatitis, or polio organisms.

Sources of Pollution

Industrial: The amount of industrial pollutants discharged into United States surface waters is about three times the amount of domestic wastes being discharged into sewers and rivers (Fig. 2.16). Additionally, this industrial waste contains more nonbiodegradable and toxic chemicals than domestic waste does.

Figure 2.16 Water pollution from a plastic industry in Maine. (Photo by author.)

The BOD of the effluent of the pulp, paper, and paperboard industries in the United States alone exceeds that of the raw sewage of the entire U.S. population. In 1970, one paper mill located on a Georgia–Alabama river raised the BOD equivalent there to that of the untreated sewage of 200,000 people. Paper industries, which have doubled in number during the past 20 years, also release "sulfite liquor," an acid containing nonfibrous material removed from wood chips during cooking. Sulfite liquor is not toxic itself but its BOD is so high that it kills fish by suffocating them. The Columbia River has an especially high level of sulfite-liquor pollution.

The textile industry, which has become concentrated during the past few decades in the Southeast, is another large-scale polluter. One Georgia mill produces discharges with the BOD equivalent of the sewage of more than 100,000 people.

The food and kindred product industry is also a giant polluter of surface waters. The canning industry pours out great quantities of waste. In the dairy processing industry, cheese plants dump billions of pounds of acid whey into rivers and lakes, the equivalent of the annual untreated domestic sewage of some 18 million people.

The chemical industry, especially the manufacturers of lye (caustic soda) and various manufacturing industries, have released many mil-

Figure 2.17 These tailings from old mines in Oregon are a source of recurrent water pollution. (USDA Forest Service.)

Figure 2.18 Strip mines for coal, now abandoned, are a major source of water pollutants. Here polluted waters still seep from a cut abandoned years ago. (USDA Forest Service.)

lions of pounds of mercury into U.S. waterways in recent years, causing high-toxic mercury levels in fish and shellfish, the concentration rising as small fish are eaten by larger fish (and eventually by people). Mercury poisoning can damage the brain, the nervous system, and deform the unborn. Much of the mercury added to surface waters finds its way into oceans where it has been found in many fish in dangerous quantities. Some details of this newly discovered environmental threat will be considered in Chapter 4.

The coal-mining industry is a major polluter of streams. The union of air, water, and sulfur compounds associated with coal forms pools of sulfuric acid water in coal mines which are then washed into streams and lakes by heavy rains. Highly acid water coagulates the mucous covering of fish gills, halts the blood flow, and causes the fish to suffocate. Millions of U.S. acres are being smothered in mining wastes (Fig. 2.17) and thousands of miles of streams, and millions of acre-feet of lakes, are polluted by the mining industry. In Pennsylvania alone more than four billion gallons of acid drainage from active anthracite and bituminous coal mines and millions of gallons of polluted water from abandoned mines enter the state's streams each day. Abandoned strip mines, especially in Appalachia, will continue to pollute, perhaps for centuries, unless they can be sealed off from draining into streams (Fig. 2.18).

Figure 2.19 Water pollution from domestic and other sources is on the rise almost everywhere in the United States as the population, affluence, consumption, and waste disposal increase. (U.S. Department of Interior, Bureau of Land Reclamation.)

Domestic: In 1970 the sewage of over five percent of the population in the United States served by sewer systems was discharged completely untreated into waterways; that is about 1.5 billion gallons of raw sewage (Fig. 2.19). Most of the domestic waste from the urban population in this country gets only secondary sewage treatment, which is now considered minimal. Every year another 1000 communities outgrow their treatment facilities for sewage.

Detergents used for many domestic cleaning purposes are rapidly increasing contaminants of surface and ground waters in the United States. Many domestic detergents contain phosphate (phosphoric oxide), the key nutrient in promoting excess algal growth (one pound of phosphorous may support more than 80 pounds of algae), resulting in eutrophication which eventually kills aquatic life by depleting the oxygen content. Phosphorus was foremost among dangerous water pollutants in 1972.

Agricultural: Agricultural runoff pollutes water with minerals, sediment, fertilizer, and pesticides. Fertilizers being used widely are the source of nitrogen, phosphorous, potassium, calcium, and other elements. Nitrogen and phosphorous (especially in nitrate and phosphate forms) cause eutrophication. Nitrate forms nitrite which can be toxic to cattle and injurious to babies, causing methemoglobinemia. Nitrates are very mobile and enter both ground and surface waters, whereas most phosphorous and potassium are bound to clay and enter surface waters and stream sediment.

About two billion tons of waste ($\frac{1}{3}$ of it liquid) from livestock and poultry are produced annually in the United States (Fig. 2.20). This is equivalent to 10 times the human population of the United States. A feedlot of 10,000 cattle is equivalent to a city of 164,000 people in sewage disposal problems. These wastes can cause eutrophication of lakes, fish kills, and nitrate contamination of aquifers. The BOD of these wastes is high: from manure-holding tanks it may be 100 times as concentrated as that of municipal sewage; from feedlot runoff it may be between 10 to 100 times as concentrated as that of municipal sewage.

In addition to the nitrates polluting water through runoff, another kind occurs: nitrogen pollution of surface waters by absorption of ammonia volatilized from animal feedlots. In addition to causing eutrophication, even low concentrations of ammonia absorbed by alkaline water may be extremely toxic to fish and other life.

Polluted Rivers

The "top 10" of United States dirty rivers in 1970 were, in order: Ohio River; Houston Ship Channel; Cuyahoga River, Ohio; River Rouge, Michigan; Buffalo River, New York; Passaic River, New Jersey; Arthur Kill River, near New York City; Merrimack River, New Hampshire and Massachusetts; Androscoggin River, Maine; Escambia River, Alabama and Florida. The 10 runners-up included the Hudson, Mississippi, Potomac, Missouri, and Connecticut River.

Figure 2.20 Water pollution from cattle feed lots is increasing rapidly as this profitable but inhumane method of cattle raising becomes ever more popular in the United States. (USDA Soil Conservation Service.)

The vital Ohio River serves as a major example of failure in cleanup efforts. In 1970 after the largest river cleanup effort in the country, with $1 billion spent in over 20 years, the Ohio was still the nation's dirtiest river. A good part of the cleaning effort was offset by the Army Corps of Engineers improving it for navigation, and essentially making the river into a series of long ponds with little flushing capacity. This allowed waste and heat to build up, combined with low aeration. The Ohio has a 10-state drainage area with a population of 24 million, and some 38,000 industrial plants along its course.

Water pollutants in the Ohio River include sediment loads, residual organics and nutrients, acid drainage, and chlorides from industrial operations. The Ohio region produced $\frac{3}{4}$ of the United States bituminous coal and had $\frac{2}{3}$ of the total acid mine production. A tributary, the Monongahela River, flows through coal country for 129 miles to Pittsburgh where it empties an annual 200,000 tons of sulfuric acid into the Ohio, contaminating it for 170 miles below the junction. Already at several locations along the Ohio, the BOD is so great that the dissolved oxygen drops to zero for extended periods, killing local aquatic life.

Polluted Lakes

Lakes are formed where the flow of rivers has been blocked and the water has been backed up behind dams or other constrictions. They are natural catch basins for sediment and often for various polluting substances as well. Very few lakes in the United States are still in their original state. Most have turned gray and murky as they were made to serve the many needs of the increasing population. The larger the lake, the more pollution it can hold, the more numerous the municipalities and industries along its banks and the more difficult the problem of regulating their waste discharges.

Our largest lakes of all, the Great Lakes, forming an interconnected group, are presently the focus of nationwide attention because of their rapidly deteriorating condition. The state of each of the five Great Lakes depends on where it is in the drainage chain and on its particular concentration of population and industry.

Erie is the most infamous of our Great Lakes. Hundreds of newspaper articles in recent years attest to the wide interest in its deteriorating condition. Into Lake Erie drain the waters of all the other Great Lakes except for Ontario and, with them, the industrial, agricultural, and domestic wastes of at least half a dozen states. The list of different kinds of chemicals, oils, pesticides, and fertilizer components to be found in Erie is endless. Long past the stage of a polluted lake, Erie is much more a chemical tank from which life is rapidly disappearing.

Seventy-eight large industries around Lake Michigan or on tributary rivers are discoloring its water with acids, steel mill scalings, oil, and pulp paper dregs. The lake is now threatened by early stages of eutrophication, being caused largely by the 177 lakeside municipalities dumping tons of sewage phosphates into the lake. The phosphates cause abnormal algal growths that are damaging the beaches and clouding the water. Of the cities, Milwaukee is the largest polluter, with South Bend, Indiana, second (dumping into the St. Joseph River, a Lake Michigan tributary). Chicago's rich north shore dumps enough pollutants into the lake to require periodic closing of municipal beaches.

Unfortunately, the lake has a very low water renewal rate, and high retention. It flushes out completely only in hundreds or even thousands of years. And the population of the Lake Michigan region is rising rapidly.

Ground Water Pollution

For most of his history, man has deposited his wastes on and in the ground. In the "underdeveloped" regions of the world, this is still the common practice, but in the United States, the largest part of the population is provided with sewer systems to carry its wastes, treated or untreated, into surface waters. In many suburban and most rural regions, however, wastes are still entering the ground, being discharged into septic tanks or cesspools.

The percentage of industrial waste discharged into underground disposal systems is small compared to the common practice of dumping by factories into rivers and other surface waters. Nevertheless, the total quantity of industrial sewage and waste materials discharged below the ground is increasing rapidly. Industrial wastes are now being disposed of by discharge into thousands of wells throughout the country.

In general, soils and other porous layers below the ground appear to be able to cleanse large amounts of water effectively of bacteria and many other pollutants by straining, absorption, and biological degradation. But such ground materials have only limited selective ability to remove the chemical compounds contained in most industrial waste.

Another source of ground water pollution is cropland. Nitrates in fertilizers are very mobile, being readily leached from soil and carried down below the water table.

In California's San Joaquin Valley, large ground water reserves are being threatened with nitrate pollution from poor irrigation practices. This valley contains eight million acres of the most fertile land in the United States and uses enormous amounts of nitrate fertilizer. The nitrate content of the valley's ground water has been rising for years.

In regions of rapid growth of suburban regions, crowded communities often develop without central sewer systems, each home dependent on its small piece of ground for both water supply and waste disposal. It is not unusual for wells in such areas to yield water contaminated with pollutants derived from neighborhood sewage. Detergents are often the first contaminating substances to be detected in the wells because of their foaming action.

The use of biodegradable detergents that break down through natural processes has been growing in recent years. Unfortunately, even biodegradable detergents pose a problem in crowded communities that discharge their wastes into the ground and also depend on ground water for home use. Under these conditions, there is insufficient time for the detergents to decompose.

In New York's Suffolk County over one million persons deposit sewage into the ground, mostly in cesspools, and detergents have been showing up in the wells which pump water from below the surface. In 1970, in desperation, Suffolk County became the first part of the United States to legally ban the sale of detergents.

Detergent contamination, in a moderate degree, may be more annoying than dangerous, but there have been thousands of cases of disease and poisoning caused by other pollutants in well waters in the United States. Gastroenteritis is the most frequent malady associated with ground water pollution, but shigellosis, diarrhea, and typhoid fever are also included in these known instances of contamination. Recent studies suggest that some malformations of infants may be a result of chemical contamination of water consumed by pregnant women.

Water Purification

When a stream of water is moving fast enough it is able to cleanse itself of a limited amount of organic waste, the amount being dependent on the rate of flow, amount of water involved, and other factors. Natural aeration of the water adds oxygen as it flows. The oxygen permits degradation of the waste and purification of the water. This has long been the natural manner in which wastes were purified and eliminated. But as the population has increased, the rivers have not increased their flow, which is now inadequate to cope with the wastes being discharged.

The ancient sewage system of Rome consisted of the Tiber River water being redirected so as to flow through the drains of the city. The quantity of water was enough to greatly dilute the waste material produced by the citizens. The effluent that left the city was probably inoffensive because of the dilution, and the bacterial action, and the oxidation that took place in the running water. Thus, the drainage into the Tiber was essentially harmless, and the river was left unchanged.

One of the simplest ways artificially to treat water for domestic use, once it is polluted, is to store it for a time in open reservoirs or tanks where sunlight can act on its surface and *aeration* takes place. Spraying water from jets or allowing it to cascade will accelerate the aeration process and the oxidation of dangerous bacteria (Fig. 2.21). Frequently an initial step in water treatment is *screening* to eliminate floating or suspended material, with a next step taking place in holding basins where sedimentation of suspended particles takes place. Sedimentation may be made more effective by the addition of a substance like aluminum hydroxide, the ions of which attract suspended particles and form spongy clusters which settle out more rapidly. This is called *flocculation*.

Most water treatment plans today depend heavily on the process of sand *filtration*. This is a filtration process in which water sinks through a sand bed. In a simple sand filter, the upper part of the bed, usually covered by a screening film of dirt, holds minute forms of plant life that are active in the decomposition of the organic matter, releasing oxygen, an important component in the purification process. Below this thin upper layer is a thicker zone in which nonpathogenic bacteria abound and complete the decomposition of organic matter in the water.

Simple sand filters, as multipurpose cleansing devices, are surprisingly effective means of purifying polluted water, but the process is a slow one. The action can be accelerated as much as 20 times by using a rapid sand filter in which water moves more rapidly, and the upper dirt screen and plant life zone does not develop; but heavily polluted water may not be fully purified. Slow sand filtration methods, for example, may

Figure 2.21 Aeration is an important purification technique in the purification of urban water supplies.

remove over half of the synthetic detergent content of water, but rapid sand filtration has little effect on the chemicals. Biodegradable detergents may be entirely broken down and removed by the biochemical action of slow sand filters.

Chemical Purification: Sterilization of water by means of chlorine or other bacteria-killing agents becomes necessary where the water available for domestic use has a high level of organic pollution. The fact that bacterial life is inhibited by low concentrations of chlorine has made this element invaluable to municipalities for drinking water purification. Ozone, another chemical used for water sterilization, eliminates bacterial life by thorough oxidation. Ozone is more effective than chlorine, but its use is relatively expensive. Thus, chlorine remains the most important chemical for water purification. Chlorine itself is essentially tasteless but by-product compounds may be produced which, though harmless, are unpalatable; their taste has become increasingly familiar to urban dwellers.

Complex Treatment: We have seen that the treatment of waste water, in order to render it harmless or minimize the degree of damage to the receiving waters into which it is discharged, may involve a number of different cleaning processes. These methods are usually described as parts of two broad classes: primary treatment and secondary treatment.

Figure 2.22 Biologic filtration stage in water purification of the Sacramento water supply. (U.S. Department of Interior, Bureau of Land Reclamation.)

Primary treatment usually consists of essentially nonbiological procedures such as screening, flocculation, and sedimentation. For mildly polluted water, this may be sufficient for purification.

Secondary treatment usually consists of biological oxidation procedures that copy natural biological assimilation and degradation processes (Fig. 2.22). The most popular type of secondary treatment is the activated sludge method. The details and variations of this procedure are extremely complex. In general, the waste water is well-aerated in tanks to provide oxygen for degradation of organic matter by microorganisms. Solid produced by the degradation are removed by sedimentation, and the clarified effluent is discharged. Some of the solids (sludge) are returned to the aeration tank; the remainder must be removed, carried off, and disposed of (one of the most costly parts of the operation). Another popular method of secondary treatment is the *trickling filter* that uses beds of rock or other objects on the surfaces of which the tricking waste waters undergo biological degradation.

In 1970 about 10 percent of the waste waters from communities with sewers still went entirely untreated. About 30 percent received only primary treatment, and only 60 percent of the waste waters received the secondary treatment usually necessary to render it relatively harmless. The name *tertiary treatment* is sometimes used for additional advanced procedures that may allow waste water to be cleansed to the level at which it can actually be reused directly by a municipality. These advanced procedures may include microscreening, chemical coagulation, and addition of bacteria-inhibiting or killing chemicals, such as chlorine and ozone.

INCREASING THE WATER SUPPLY

The history of our water supply as the population increased and per capital needs grew, has consisted chiefly of reaching out farther or digging deeper. We have so nearly reached the limits of present supply that if we are to continue to grow in numbers and consumption, it will be necessary to: find still more considerable, still unused supplies; develop techniques of producing additional freshwater, in effect, altering the hydrologic cycle; or we must reuse large amounts of previously used or waste waters.

Increasing Rainfall

The technique of seeding clouds in order to initiate or increase rainfall was developed over two decades ago. Clouds passing over the United States each day are estimated to contain over six times as much water as the total precipitation that actually falls. Some of this cloud water may be brought down as rain by seeding processes in which fine particles of carbon or silver iodide are spread from plants or from ground apparatus. Most of the commercial seeding operations carried out thus far have been from planes and were extremely costly emergency measures over drought-struck, high-value cropland.

Where cost is not a controlling factor, it is theoretically possible to increase rainfall by as much as 20 percent over broad regions by regular seeding operations. But in a practical sense, increasing rainfall by seeding techniques must, for the present, be considered too costly and results too uncertain to be used on more than a very small and local scale.

New Dam Construction

The irregular distribution of rainfall during each year presents a basic problem of water supply that has been met partially by building dams to hold back the runoff during the times of heavy rainfall, that is to be used in drier months. Dams are also important for flood control and form reservoirs used for irrigation (see Chap. 5), recreation, and hydroelectricity (see Chap. 3).

The United States now has about 1500 major reservoirs and many smaller ones. The construction of new dams for additional water supply has been meeting increasing opposition from conservationists distressed by the prospect of more valleys being submerged and their scenery forever lost. Just prior to this century, the opposition by a group of people to San Francisco's plans to build a dam and inundate a deep, scenic valley just north of famed Yosemite Valley in California, was instrumental in the birth of the nation's chief conservation organization, the Sierra Club. The group's fight was unsuccessful, and San Francisco gained a major reservoir for water supply but, during the following 70 years, the opposition to dams has been steadily stiffening.

One of the arguments against the proliferation of dams is the rate at which the reservoirs tend to silt up, especially in regions of dry climate and vegetation too sparse to hold back runoff after the periodic rains (Fig. 3.7). The large Gibraltar reservoir that originally provided the water supply for Santa Barbara, California, is a good example of the impermanence of reservoirs. This reservoir was built in 1920 and filled with silt so rapidly that in 25 years its capacity had shrunk by half (Fig. 2.23). Two secondary dams, built upstream to stop the sedimentation process, had their reservoirs completely filled with silt in a few years (Fig. 2.9). An additional major reservoir was subsequently constructed downstream from the Gibraltar dam. But the silt is still settling, and the population of the region still increasing.

Figure 2.23 Gibralter Reservoir, California is being rapidly filled with sediment, thereby seriously decreasing the usefulness of this artificial lake for water supply purposes. (USDA Forest Service.)

Evaporation Losses

Enormous amounts of freshwater are lost each year by evaporation from the surface of water bodies. Daily evaporation from lakes and reservoirs in the arid western states is over 10 billion gallons per day! Each year, approximately seven feet of water are lost by evaporation from enormous Lake Mead behind Hoover Dam. Many lakes and reservoirs in the Southwest have lost even more in a single year of evaporation.

Recent research shows that evaporation rates can be decreased by blanketing lakes and other water bodies with thin protective layers of chemical substances. Experiments with sprays yielding monomolecular (one molecule thick) films have been carried out in a number of regions in attempts to significantly diminish evaporation rates. It may be possible to reduce evaporation losses by 20 percent or more by such protective films, and this would be of greatest value in arid regions. But there are many problems involved, including the tendency for rain and wind to break up or remove the film.

Evaporation losses during agricultural irrigation are enormous, more so because most irrigation is carried out in regions of low humidity and rapid evaporation. In some regions, as much as $\frac{2}{3}$ of the water being used for irrigation may disappear through a combination of evaporation and seepage into the ground. Some of these losses could be reduced by lining irrigation ditches with concrete, or other material, or by conveying irrigation waters wherever possible by means of closed conduits.

Much water that would otherwise be used by man for domestic water supply or irrigation is being evaporated (transpired) through the action of trees and other vegetation. In regions where the water supply is especially critical, there have been recent attempts to reduce evaporation losses by defoliation or other means of destruction of phreatophyte plants (plants capable of transpiring large amounts of moisture from soil) in watershed areas, thereby permitting downward percolation of a greater fraction of rainwater and increasing the flow of water in the rivers draining the regions. But there are adverse side effects to watershed defoliation that include the loss of topsoil by rain after the vegetation is gone and damage to animal and even human life from the toxic chemicals used.

Desalinization

Desalinization is a short circuit of the hydrologic cycle in which sea water is artificially changed into usable fresh water. Many desalinization plants are now being built or planned. But the cost of the desalinization water, while only a fraction of what it was 20 years ago, was still about $1.00 per thousand gallons in 1970. This was many times the cost of freshwater in this country even in the arid Southwest where the race rarely exceeded $0.25 per gallon. But it is expected that technological advances will eventually reduce the cost of desalinized water considerably.

Desalinization can be accomplished in a number of ways. These include the familiar distillation process, requiring boiling and condensation; the crystallization process in which freezing is used to separate salt from water; the membrane process involving diffusion through semipermeable membranes to achieve the separation; electrodialysis, in which electrodes remove the salt ions; and a number of more complex and experimental means and variations.

Many countries are placing their hopes for growing, thirsty populations on the prospect of large-scale desalinization. Indeed, there are already hundreds of small, working desalinization plants in the world, including a number in the United States. Some of the dry Eastern Mediteranean countries, especially, are increasing their desalinization capabilities. Nevertheless, in 1970 all of the world's desalting operations produced less than 100 million gallons of water a day which is about a tenth part of the daily water consumption of the city of New York! The most optimistic estimates of the world's desalinization output for 1984 is about $\frac{1}{30}$ of the world's estimated total needs during that year. Even if a large supply of water is made available by desalinization, it will probably be much too expensive to use for irrigation and certainly too expensive if it must be pumped inland for any distance (see Chap. 5).

Search for More Water

Water From the North: Among the suggestions made for finding new sources of water has been the scheme to divert much of the waters of Alaska into channels or giant aqueducts that would carry it southward to population centers in California and elsewhere. But it has been recognized that the removal of vast quantities of water from Alaska would have major, perhaps catastrophic, effects on the local water tables and on the water supply for all plant and animal life. The ecological disruptions would seem to be endless.

In addition to the local havoc that would be wrought, the mechanics of moving amounts of Alaskan water as immense as some have visualized are probably beyond present technology. Much of the terrain is mountainous; the distance is some 1500 miles and the costs involved are so staggering as to render the Alaskan venture impractical for the foreseeable future.

Perhaps still more impractical is the recent suggestion that huge icebergs from Alaskan or Greenland glaciers be towed southward to be melted near major population centers. The limited number of very large icebergs that break free each year and the high cost of towing these irregular, melting masses of ice a thousand miles or so are only two of the factors that place this scheme far from the realm of realistic answers to the water supply problem.

Still Untapped Ground Water

Although water is to be found under all land surfaces of the world, its quantity and availability is extremely variable and often very limited when compared to the needs of the populace living above. Nevertheless, there are vast supplies of still untapped ground water in certain parts of the world. A prime example consists of the thick, cavity-riddled formations of limestones that underlie much of Florida and sections of adjacent states. This giant, many-layered aquifer presently feeds a number of artesian wells but is still essentially untapped and probably will be of greater future importance to the growing population of the east coast.

The state of Nevada that has about the driest climate of the United

States has been shown to have great quantities of ground water in its numerous low-lying basins. Deep below the sage-covered floors of these mountain-bounded depressions are piles of porous sediment that have been slowly filled with water by many years of downward-soaking waters after periodic heavy rains. Large-scale pumping of these waters holds great promise for irrigation of some of the arid basins and for other uses in Nevada. But the use of underground water that has accumulated so very slowly, sometimes called fossil water, is similar to placer mining in which mineral deposits laid down over thousands of years by slow stream processes are mined out in a few years. It is very likely that once pumping has begun, it will be carried out so rapidly that the water will be exhausted, essentially mined out, in a matter of a few tens of years at most. Refilling of the basin strata would eventually ocur but at a rate hundreds of times slower than it is likely to be pumped out.

The thick strata of the coastal plains in the Southeast and the blanket-like sediments of glacial origin located in a number of Northern states are among examples of regions of major, proven underground water reserves in the United States. In every case, there exists the danger, once the aquifers are tapped, of too rapid a withdrawal and subsequent depletion by the unregulated exploiter, which is all too frequently the case in the United States.

Submarine Springs

A large proportion of the ground water under near-coastal regions eventually seeps out into the oceans, mingles with salt water and is lost to man as a source of freshwater. In a few parts of the world, major underground rivers gush out below sea level in such a concentration that the freshwater comes to the surface offshore still uncontaminated by brine before diffusing into the surrounding sea. Submarine springs have been known for thousands of years off the coast of Greece near Taranto, and Syracuse, and for nearly half a century near St. Augustine in Florida. It is probable that there are thousands of places where fresh ground water issues from the submarine edges of land areas undetected because of deep water or because the flow is too weak to reach the surface before mixing with the salty sea.

It has been suggested that major submarine springs, like the one off St. Augustine, could be used for water supply by surrounding the rising waters with a floating barrier and carrying them by floating or fixed aqueducts to pumping facilities on nearby land. Submarine springs are unlikely to be of major importance to man unless many more springs, presently undetected, were to be located and their waters directed back to the land and its population centers.

Recycling Water

An important consideration in evaluating the quantity of water available for use in a region is whether or not the water is to be used in such a way that it will be largely returned to its source after serving its function, or whether a large proportion of it will be lost, usually by evaporation, during its use. Most of the water drawn up from wells or streams for domestic use, for example, is discharged back to a water body after use. Almost all of the water taken from a home well is usually returned to the water table in the same vicinity after passing through a septic tank or apparatus of similar purpose. By contrast, more than half of the water used for irrigation is commonly lost by evaporation, usually seriously depleting the quantity of ground water available in the region.

In the United States the total natural runoff (amount of water which would appear in streams if there were no artificial developments) is about 1200 billion gallons a day. By 2000 A.D. it has been estimated that $\frac{3}{4}$ of this amount will be needed for domestic agricultural and industrial use. Considering the distribution of water sources and of population and industrial centers, it is doubtful that such an amount of water could ever be obtained without reusing local water a number of times.

In some regions where increasing demands for ground water have seriously lowered the water table, procedures have been established to use surplus water from adjacent regions or to reuse local waste waters in order to recharge the underground zone of saturation and raise the water table to its former level.

Recharging may mean that water is spread over a land surface (or into hollows) from which it can seep downward to add to the ground water and raise the depressed water table. Water may also be injected into wells, thus, reversing the common process and raising the water table around the wells. Surface waters available in excess in one region may be used to replenish a water table in a nearby region that has fallen because of severe pumping withdrawals by local users. Or surface waters may be impounded in reservoirs after floods and later spread out to replenish ground water, instead of allowing most of it to escape rapidly through streams after the heavy rains.

Recharging with waste water is a form of recycling water, the waste or sewage water being reused after the natural filtration of percolating through porous ground. This use of effluent seems to be safe when done carefully because most pollutants, including coliform bacteria, will be removed by soaking a relatively short distance through fine-grained sediments. It is a method of enormous potential when we consider the rate at which most of the very municipalities that so drastically deplete their ground waters are producing sewage.

At present, the largest amount of recycled water is that being used for process cooling by industry. Overall water use by industry averages two gallons for each gallon of new water withdrawn. Reuse of water by American industry is expected to increase drastically in the coming years. In stream and electric power production, the increasing use of cooling towers (Fig. 3.20) should allow the withdrawal of less water for cooling purposes from surface water bodies but may mean the loss of larger proportions of that water through evaporation (see Chap. 3).

Future Water

If near-elimination of water pollution is ever achieved, more effective means of water storage and distribution are developed, and recycling procedures, especially by industry, are stepped up, there should be enough water for the present population level at least for a few decades. This would only apply in the arid regions if major engineering works are constructed to channel water from better watered regions or, in the case of coastal regions, huge-scale desalinization works are established. Watering the arid regions by desalinization will require vast expenditures, and possible subsidizations by the population of the humid regions, and involve serious environmental hazards considered in Chapter 3 because of the massive production of electricity needed. Yet, it is certain that water supply will become increasingly critical as the world's population rises. It is not unlikely that without major desalinization works water will become completely inadequate for man's needs, especially for increasing his food supply, as his numbers increase.

Suggested Readings

American Chemical Society, *Cleaning Our Environment: The Chemical Basis for Action,* a report of the A.C.S., 1969.

KAZMANN, R. G., *Modern Hydrology,* Harper and Row, New York, 1965.

LEAR, JOHN, "The crisis in water, What brought it on?" *Saturday Review,* Oct. 23, 1965.

MACKENTHUM, K. M., *The Practice of Water Pollution Biology,* Federal Water Pollution Control Administration, U.S. Government Printing Office, 1969.

NACE, E. R., *Are we running out of water?,* Geological Survey Circular 536, 1967.

NACE, R. L., *Water and Man: A World View,* UNESCO, 1969.

PIPER, A. M., *Has the United States Enough Water?,* U.S. Geological Survey, Water Supply Paper, 1797, 1965.

STEGNER, WALLACE, "Myths of the western dam," *Saturday Review,* Oct. 23, 1965.

3
Energy

Photo courtesy of Bureau of Reclamation, Department of the Interior.

Energy is the capacity to perform work, but it comes in many forms. Man is unique among animals in the degree to which he concentrates large amounts of energy from his environment in order to decrease the amount of his own labor and yet raise the quality of his existence.

Potential energy is an energy of position. Water impounded behind a dam has the potential to turn water wheels or turbines when it is finally allowed to fall to lower levels. *Kinetic energy* is to be found in moving bodies. The work may occur when the movement is arrested. Thus, the wind posesses kinetic energy that was used by means of the old windmill, the work of which was grinding flour or pumping water. Heat is *thermal energy*. The heat of a flame performs the work of boiling water, cooking, making steam for heating houses, or turning electric generators. The heat of the sun is the most vital form of energy in our environment, energy without which no life could exist.

When primitive man discovered the use of fire to keep himself warm, he took the first great step on the road to an industrialized civilization with its massive network of energy-consuming devices. As civilization developed, his use of energy was extended to the harnessing of wind and running water by means of windmills and waterwheels. He learned to smelt metals by using wood fires and wood-derived charcoal.

A giant step in energy utilization came early in the 18th century when Europeans began to use coal. Coal became a major factor in the development of the steam engine that was used to drain water from coal and iron mines as they were sunk ever-deeper into the ground. Steam power, produced first by burning wood and then by coal, was soon used for transportation (locomotives) and many manufacturing purposes in the Industrial Revolution of the last century.

In the last decades of the last century, petroleum began its development as the major competitor of coal as a source of concentrated energy for heating and transportation. The development of the internal combustion engine and its need for gasoline or other derivatives of petroleum was a major factor in the explosive development of the petroleum industry during the past half century.

The past 30 years have seen the birth and early development of the newest and most fabulous source of power yet, the energy released from the nucleus of the atom itself.

In 1971 about 95 percent of the energy used in the United States came from the combustion of fossil fuels or their derivatives (Fig. 3.1). Only about four percent of the energy came from falling water (hydroelectricity) and a little more than one percent came from atomic power plants. The three principal fossil fuels: coal, petroleum, and natural gas were used in roughly equal amounts. Although energy was used in these various forms, our heavily-industrialized nation drank up energy at a rate equivalent to well over 2500 gallons of gasoline per individual during 1971.

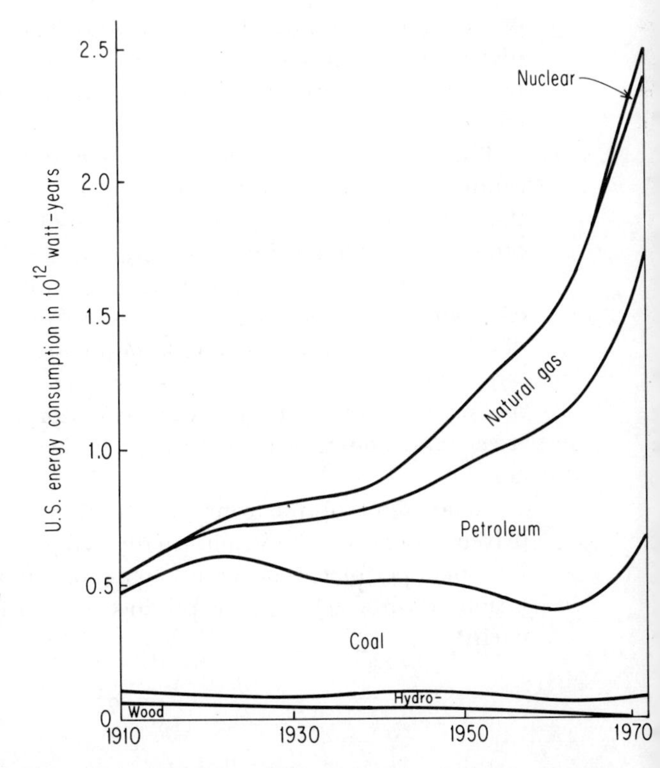

Figure 3.1 Growth trends in the sources of energy in the United States from 1910 to early 1972.

Figure 3.2 In the United States most electricity is produced by the burning of fossil fuels. New Mexico coal-burning power plant. See also Figure 3.26. (Photo by author.)

AN ELECTRIC NATION

Growth of Electricity: One of the most extraordinary developments in the history of American energy use is the increasing reliance on electricity. Electrical power is converted from another energy source and conveniently transmitted through wires (Fig. 3.2). It is now the most popular form of energy used in American homes and vital to modern industry. Few Americans ever go more than a few hours without using electricity.

The earliest forms of electrical current were produced in the mid-19th century by voltaic batteries made up of copper and zinc plates. Late in that century, as a result of the inspired experiments of Faraday and others, it became possible to generate in a wire a continuous flow of electricity by the use of magnet-bearing generators turned either by means of steam or falling water. It was the subsequent work of Edison who developed the electric-powered incandescent light, that led to the first large-scale use of electricity. This occurred in 1882 when Edison closed the switches on a steam-driven generator in New York City, and the streets and houses were bathed in the brillant light of the new electric era.

Today most of the country's electricity is still produced by steam-driven generators, the steam usually being produced by the heat of burning coal, petroleum or natural gas, or, in a few instances, of atomic fission. Hydroelectric plants produce a small fraction of the nation's electricity.

The consumption of electricity in the United States is rising rapidly. In 1971 the annual consumption was about 7000 kilowatt-hours per capita (one killowatt-hour is the energy used by a 100-watt light bulb burning for 10 hours). This is about $3\frac{1}{2}$ times as great as the per capita consumption in 1950.

Inefficiency of Electricity: Electricity allows the separation of the powered appliance from the source of the power. A major example of this advantage of use is electrical space heating that has become popular in place of oil furnaces with their need for fuel deliveries, installation spaces, chimneys, and fumes. But the convenience is dearly bought. *Electricity is the most inefficient of the common forms of energy used today as well as the most expensive in both environmental and monetary costs.*

Most of the electricity generated in this country today is produced by burning coal or other fossil fuels (Fig. 3.3). But the waste is enormous. The loss of energy that occurs during the generation of electricity in addition to the amount lost during transmission through wires and in the electrical appliance itself means that more fossil fuels must be burned than would have been required had the appliance been operated by a fossil fuel directly.

Figure 3.3 Generation of electricity by fossil-fuel burning always produces environmental pollution. Power-plant smokestacks and transmission lines near Denver. (U.S. Department of Interior, Bureau of Land Reclamation.)

A good oil- or gas-burning furnace is at least twice as efficient as the average electric space heater in terms of the fuel that must be ultimately consumed. In spite of this, the use of electricity for space heating is growing rapidly, promoted by the advertising of utility companies and encouraged by the lower rates set for heavy users of electricity.

In 1971 about 10 percent of the country's useful work was being done by electricity. But producing that much electricity accounted for more than 25 percent of the gross consumption of energy because of the built-in inefficiency of electricity.

Storing Electricity: Unlike oil or coal, electricity is not directly storable in the vast quantities required by urban regions during times of

Figure 3.4 Pump storage projects are a means of storing electricity in the form of the potential energy of water. (a) During times of low need for electricity, usually night time, a city's generating facilities are used to provide electrical power to pump water into a storage reservoir. (b) During times of high (peak) need for electricity, generators are turned by water flowing from the reservoir.

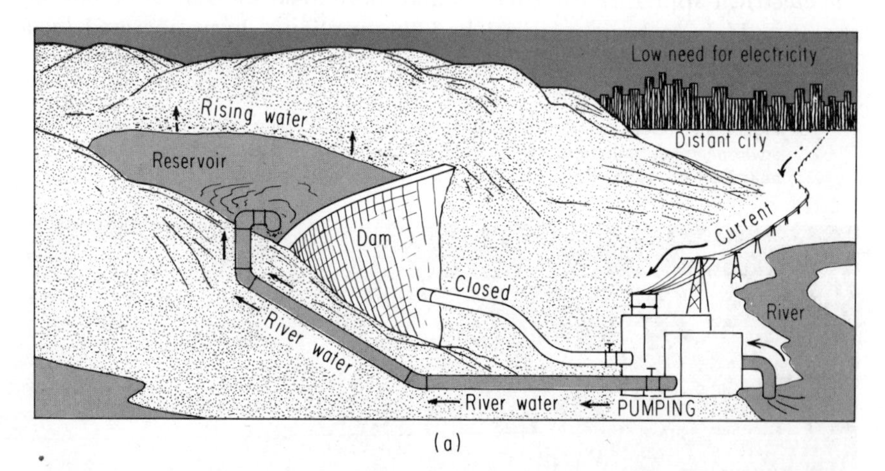

(a)

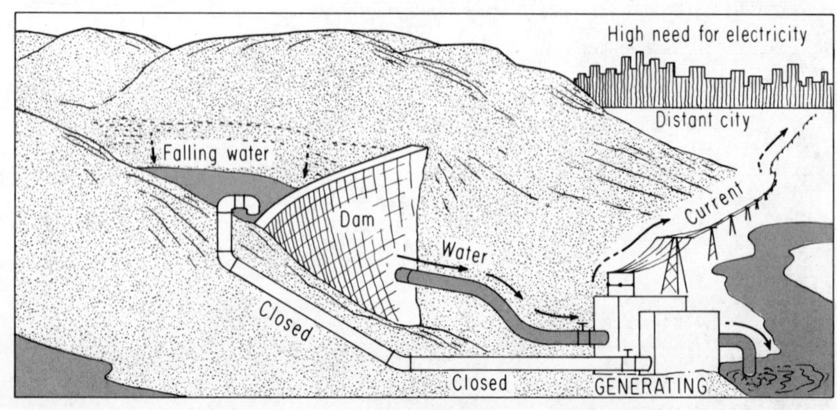

(b)

peak demand. Storage batteries have small capacity and are too expensive. One storage procedure recently developed by utility companies consists of building artificial lakes or reservoirs into which water can be pumped electrically and from which energy can later be retrieved when needed, by reversing the flow of water and letting the pump motors act as generators of electricity (Fig. 3.4).

Reservoirs of this type have already been constructed and used in a few locations, but many large-scale pump storage proposals are opposed by conservationists because of environmental and scenic damage. One concern is the possibility of causing earthquakes or tremors by rapid and frequent changes in the concentrations of weight on the earth's crust during the various stages of filling and emptying the reservoir.

A major pump-storage facility has been proposed at Storm King, about 50 miles north of New York City, in an attempt to ensure a reserve of electricity to supply New York during hours of peak demand. Here the picture is complicated by the proximity of the city's main aqueduct to the proposed storage reservoir and the possibility of contamination of the water supply by the polluted Hudson River water that is to be pumped into the storage reservoir. This region lies adjacent to a geological fault zone of considerable potential for earth movements and water leakage. In spite of these dangers and great opposition by environmentalists, the Storm King project may yet be built, so great is the demand for electricity in New York City.

WATER POWER

The energy of falling water, harnessed as water power, is energy indirectly derived from the sun. Solar heat causes the evaporation of water from the sea and other wet surfaces, thereby raising vast quantities of water into the atmosphere, allowing the formation of clouds and the rainfall which locally provides the flow of streams and rivers which are used for water power. This is the hydrologic cycle discussed in the last chapter.

The history of the use of water power dates from Roman times. By the fourth century A.D., efficient vertical waterwheels operated by rapidly moving river water were being used for grinding grains and limited mechanical tasks. By the 16th and 17th centuries, waterwheels were turning machinery and providing power for the rising industrialization of western Europe.

In the United States water power was extensively employed for driving grinding (grist) mills, saw mills, textile mills, and other purposes during the 18th and 19th centuries. But water power was confined to uses in

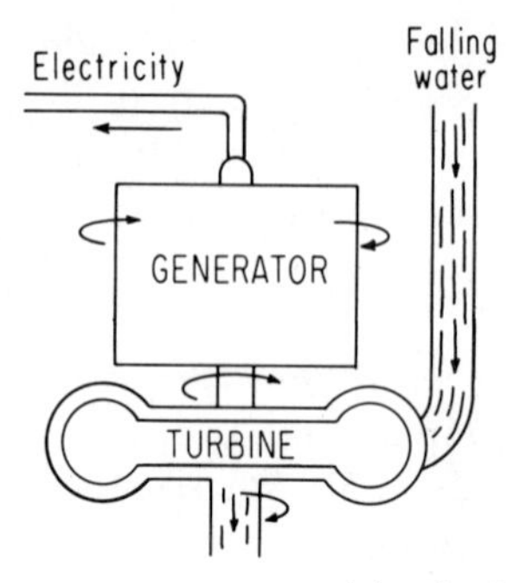

(a) Hydroelectric plant

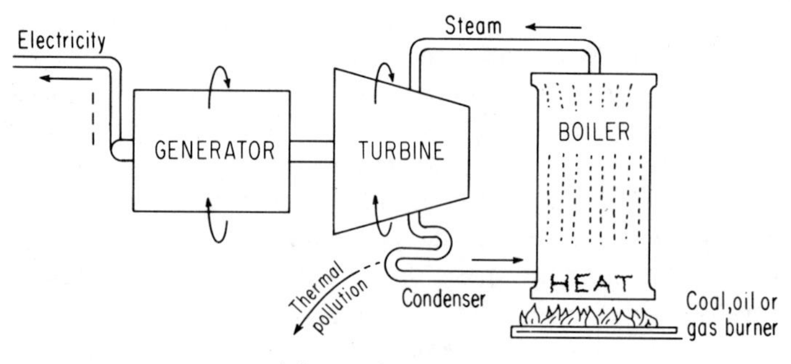

(b) Fossil fuel-steam plant

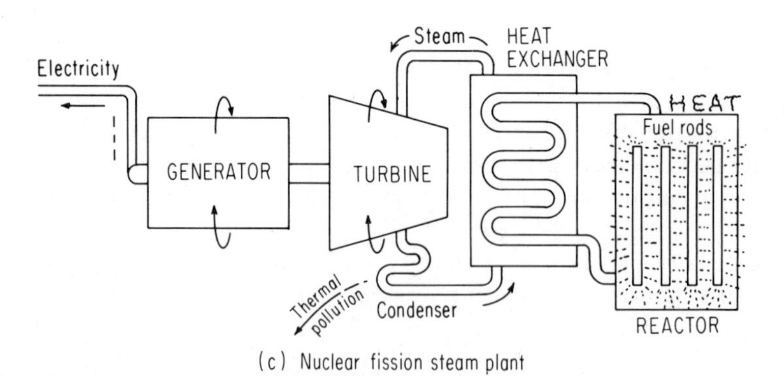

(c) Nuclear fission steam plant

Figure 3.5 The three principal methods of electric generation: (a) Hydroelectric power plant; (b) Fossil-fuel burning steam-driven power plant; (c) Nuclear fission steam-driven power plant.

the immediate vicinity of the waterwheels because of difficulties inherent in power transmission by mechanical devices. It was not until the development of electrical power transmission at the very end of the last century that large-scale and widespread use of water power became possible. Since that time, the hydroelectric power capacity in the United States has risen yearly.

Clean Energy

Using water power has many advantages. There is no fuel problem, no waste products of any kind to dispose of. Energy conversion is extremely simple; the acquired potential energy of the water is converted by the fall of water into kinetic energy, by means of the turbine into mechanical energy, and by use of a generator into electrical energy (Fig. 3.5). Modern generators achieve efficiencies over 90 percent. Control is simple and almost immediate: a hydroelectric generator can be turned on in the time it takes to open the water intake gates, often a matter of minutes. In a steam-powered plant there is considerable delay in producing the steam required to bring the generators from rest to full capacity.

Hydroelectric generators vary in size; five American water power installations have capacities in excess of 1000 megawatts each, which is comparable to the largest steam-electric generator plants; most are much smaller. The world's present hydroelectric power plants produce a total of about 50,000 megawatts, but the world's theoretical hydropower capacity is calculated to be between two and three million megawatts.

The regions having the largest but least developed water power capacities are Africa, South America, and southeast Asia, with 750 thousand, 550 thousand, and over 450 thousand megawatts each, respectively. Together they represent 60 percent of the world's *potential* but are presently developed to an extent representing less than two percent of this theoretical capacity.

Limits of Water Power

The world's water power potential is about the same magnitude as is the present world consumption of energy, most of which comes now from combustion of fossil fuels. But it is very unlikely that water power can ever approach its theoretical maximum development or replace fossil fuels to any great extent because of a number of problems and disadvantages inherent in making use of the energy of flowing water:

1. Locations of hydroelectric plants are limited to river sites where natural waterfalls occur, or valleys where an artificial fall can

(a)

(b)

Figure 3.6 Hydroelectric dams: (a) Glen Canyon, Arizona-Utah. (U.S. Department of Interior, Bureau of Land Reclamation.) (b) Fontana Dam, North Carolina. (Tennessee Valley Authority.)

be produced by the construction of high dams, and where storage of water is also feasible because of steep-walled valleys or other good sites for reservoirs (Fig. 3.6). High dams and large reservoirs are necessary to provide sufficient height of fall and to provide continuous supply of water, thereby overcoming the normal variability in flow of rivers with season and frequency of rainfall.

2. The reservoirs required for most hydroelectric installations become catch basins for mud and silt carried in by the rivers. The time required to completely fill most reservoirs is only about one or two centuries in regions of average rainfall. In very arid regions or where there have been extensive lumbering operations, the lack of protective plant cover in the reservoir's watershed region allows more rapid soil erosion; complete silting in of a reservoir may occur in a few decades or less (Fig. 2.9). Once filled by sediments, a reservoir becomes almost useless as a power source (Fig. 3.7).

In the United States there has been ever-increasing opposition to the building of new dams and the impounding of river waters for water power. In 1967 the opposition of major conservation groups, like the Sierra Club, blocked the construction of two dams proposed by the U.S. Bureau of Reclamation for construction in one of the world's most spectacular scenic treasures, the Grand Canyon of the Colorado River in Arizona.

One of the largest hydroelectric projects proposed in recent years, the Rampart Dam on the Yukon River of Alaska, would have flooded eight million acres of land. Strong opposition to this project on the basis of the killing of forests, the probability of unparalleled damage to wildlife, and other kinds of environmental destruction has probably blocked this monumental extravaganza of water power. A number of smaller dams will probably be constructed as needed in the region.

The future development of water power in the United States and most other highly developed countries will probably be quite limited by the considerations above and cannot be counted on to supply any larger fraction of energy consumption than it does at present. Indeed, many hydroelectric complexes once built, soon become surrounded by fossil fuel burning steam-electric plants because of the inadequacy of the water power as population and energy demand soar. The monumental series of dams built along the Tennessee River by the Tennessee Valley Authority have long since become inadequate for modern demands for electricity. Coal-fired power plants now have become so numerous in this region that the T.V.A. is a major owner of strip mines.

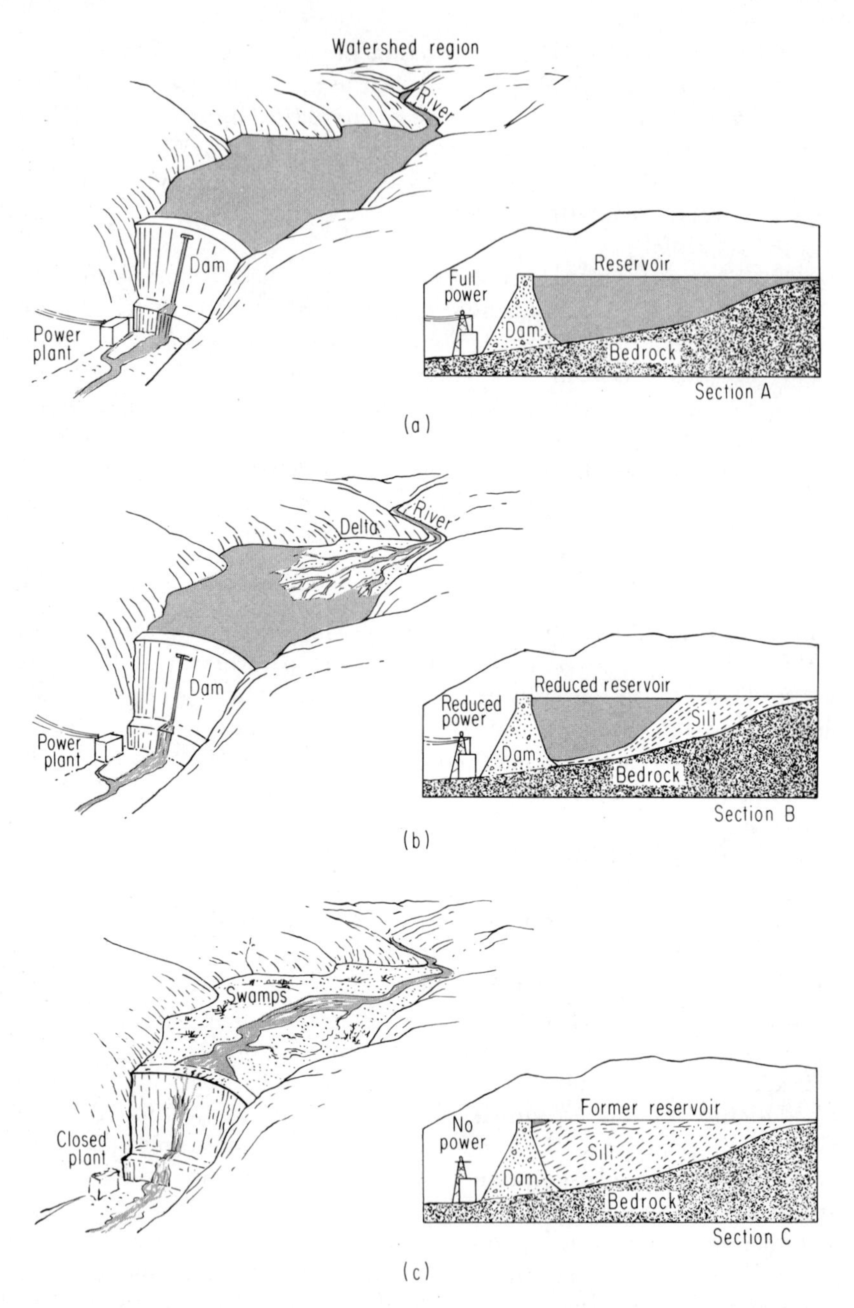

Figure 3.7 Stages in the progressive filling in of a hydroelectric reservoir by siltation.

In countries where water power potential has scarcely been tapped, future development can be expected on a large scale. This is especially true where a region has little coal or petroleum. A number of underdeveloped countries in Africa and South America are in this situation. The principal barrier to hydroelectric development in such countries most likely will be the great cost of building the dams and power plants required for water power utilization. The same problem of high initial costs will arise should the governments of underdeveloped countries choose to use atomic fission plants instead (see below).

THE FOSSIL FUELS

Millions of Years of Sunlight

Almost all of the energy we use comes from the solar radiation of the sun. Plants use this energy by the vital process of photosynthesis, and animals use it indirectly by eating plants or other creatures which depend on plants. Thus, by domesticating animals and by burning wood, man uses concentrated solar energy. If he burns a very old tree, like a redwood giant, the heat energy released represents some of the solar radiation which reached the earth's surface during the centuries that passed as the tree grew and built its thick trunk, layer by layer.

Living plants temporarily store the energy derived from solar energy; this energy is eventually released as heat to the atmosphere during the decay of the dead plants. The heat given off by decay is approximately equal to that produced and stored by the living plants so that a balance is maintained. But in any given year a very small fraction of the photosynthesized energy is held back from decay in a few unusual situations, such as in a swamp or in certain protected bodies of water.

In such environments, the lack of oxygen retards decay and allows the accumulation of organic matter and its preservation amid layers of mud or other sediments. This is the preservation process that produces the fossil fuels: coal, petroleum, and natural gas. It is quite rare, most living matter being wholly destroyed soon after death. But during the last half billion years of earth's history, enough matter has been preserved to provide us with the fuels on which we so heavily depend on today.

Man's Use of Coal

It is believed that the Chinese used small amounts of coal as long as 2000 years ago, and there is clear evidence that some of the Indians of northern Arizona burned coal prior to the arrival of Europeans. In England, coal was first used in the 12th century, having been found as

Figure 3.8 Typical coal bed in Pennsylvania. Strip mining of such a layer would require removal of all rock and soil above the coal. (Photo by author.)

scattered lumps weathering out of the rocky coastline, where they were called sea coals (or coles). For hundreds of years such coal was gathered and used for cooking and heating, but it was not until the early 18th century that it was recognized that coal occurred in extensive beds or seams interlayered with other sedimentary rocks (Fig. 3.8). In this country, coal was being mined in Virginia by the middle of the 18th century.

From the middle of the 19th century on, the use of coal increased rapidly. In the United States, usage by railroads, industry, and homes was heavy until the close of World War II. At this time, a slight leveling off of demand occurred as railroads and home furnaces stopped burning coal, but a new demand arose as water power was inadequate for the increasing demand for electricity. The use of coal by steam-electric plants soared and the rate of consumption of this fuel is now greater than ever and rising rapidly.

Fuel from the Swamps

A coal bed is a greatly compacted layer of plant matter that originally accumulated in a swamp or bog, generally along the edge of the sea. Under such stagnant conditions, the plant remains that fall to the ground year after year, century after century, are kept from decaying completely by the low oxygen content of the wet ground. Thus, a thick layer of *peat* is built up. This spongy mass of vegetal matter becomes covered by layers of mud or other sediment when a change occurs in the environment of deposition. Often the change is a result of the sea flooding the land (Fig. 3.9).

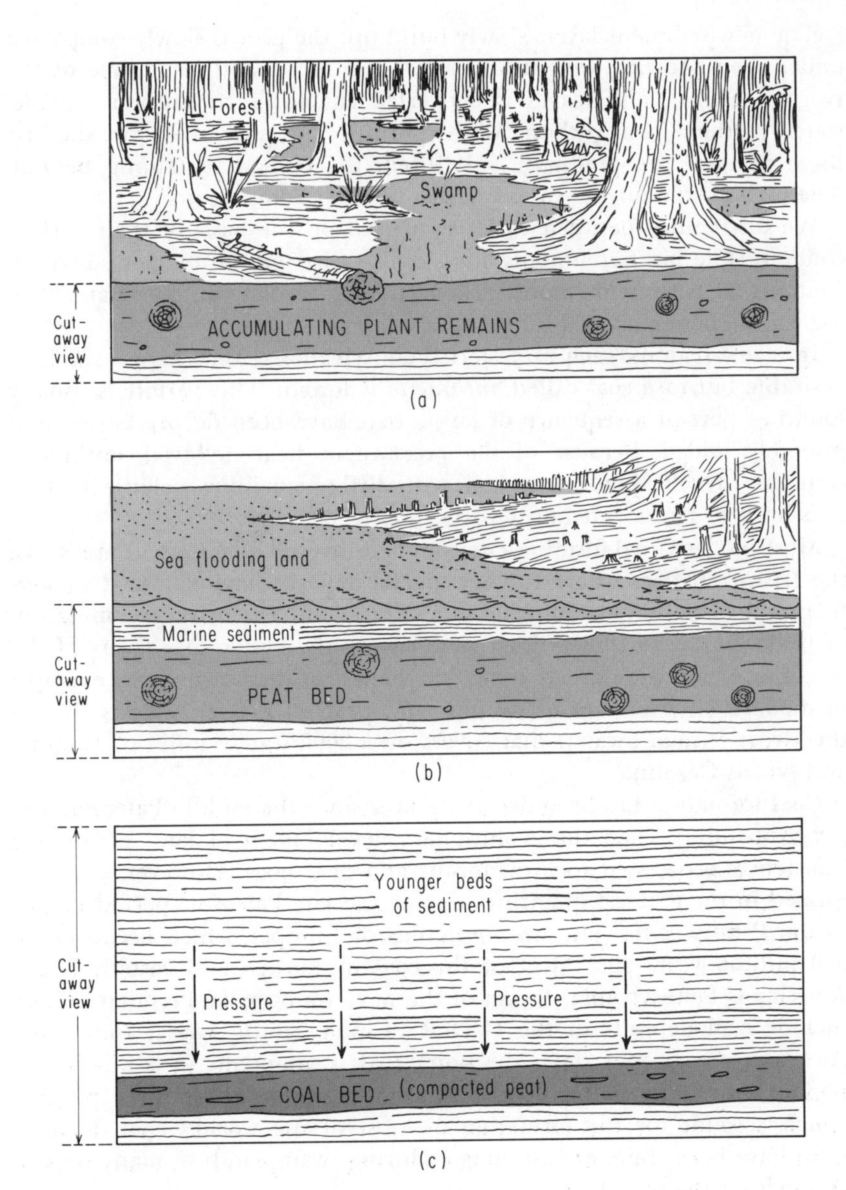

Figure 3.9 Stages in the formation of a coal bed: (a) Original swampy forest environment, (b) Flooding by sea, destruction of forest, and burial of swamp debris by sediments, (c) Compaction of swamp peat by additional younger beds of sediment to produce a coal bed.

The new sediment layers slowly build up; the peat is slowly compacted under their growing weight and more and more of the cellulose of the twigs, branches, and leaves is converted chemically to carbon dioxide, water, methane, and carbon. With continued pressure and time, the first three of these are gradually volatilized, the carbon remaining behind. The carbon content increases relative to other components.

We distinguish between varieties of coal on the basis of their carbon content. The most commonly used coal, called *bituminous* or soft coal, contains 50 percent or more carbon, greater carbon content characterizing the higher ranks of bituminous.

In a few regions, such as eastern Pennsylvania, a very hard, especially desirable but rare coal called *anthracite* is found. This variety is usually found as part of a sequence of layers that have been deeply buried and intensely folded. Because of the pressure or heat involved, anthracite contains over 95 percent carbon with little remaining volatile matter. It is hard to ignite but burns with almost no smoke.

Most modern coal-forming environments are relatively small areas like the Irish peat bogs where Ice Age glacial deposits have dammed streams to produce shallow, swampy basins. Thick and extensive accumulation of peat, similar to those which gave rise to the major coal layers of the world, are not widespread today. Perhaps, the best American example of modern environment of swamps and peat on a large scale is that of the Great Dismal Swamp that covers over 2000 square miles of Virginia and North Carolina.

Coal formation has been occurring ever since the middle Paleozoic Era (see Appendix) when the evolutionary development of land plants had sufficiently progressed to form forests and peat bogs. However, coal was formed in the greatest quantities during the late Paleozoic period known as the *Pennsylvania* (late *Carboniferous* of Europe), which began about 320 million years ago. Most of the coal in Europe and eastern North America is Pennsylvania. Strata of the next younger, or Permian period, include coal deposits in the U.S.S.R., China, South Africa, India, and Australia. Cretaceous (late Mesozoic) coal is abundant in a number of regions, especially in the Rocky Mountain states. These three periods, which account for the formation of most of the world's coal deposits, must have been times of vast, long-enduring swamplands in many regions throughout the world.

Energy in Layers

Individual coal beds (or seams) range in thickness from a fraction of an inch to several tens of feet. It has been estimated that about 30 feet of peat must be compressed in order to produce each foot of bituminous coal. Studies of rates of peat accumulation suggest that each foot of accumulation required about a century. Thus it appears that each foot of coal represents some 3000 years of plant growth. A valuable ten-foot-thick bed of coal would represent about 30,000 years of continuous plant growth during which trees and other plant debris gradually accumulated in Paleozoic or Mesozoic swamps.

It is clear from the above estimates of the time required for coal formation that *new* coal is being formed too slowly even to be considered as fuel for mankind. In practical terms, the world's present coal layers are all the coal available to us.

Geologists believe that all the major coal fields of the world are now discovered. The total recoverable amount of coal beds over one-foot thick and within working distance of the surface has been estimated at over 8000 billion tons, with Asia, North America, and Europe, in that order, accounting for over 95 percent of the total. The continents of South America, Africa, and Australia together account for the remainder, less than five percent of the world's reservoirs. Clearly, the southern hemisphere is coal-poor.

In spite of the thousands of billions of tons of coal reserves, if we continue to use coal and increase our consumption at the rates predicted by various experts, the time required to exhaust most of the coal supplies in the world is somewhere between one and four centuries. And in order for most of these reserves actually to be mined, the strip type of operation is almost certain to be employed on so vast a scale as to seriously alter the world's landscapes (Fig. 3.10).

Figure 3.10 Strip mining for coal, shown here in Colorado, causes major environmental damage. (USDA Soil Conservation Service.)

Figure 3.11 Millions of acres of American landscape, often of great scenic value or agricultural potential, is converted each year into devasted terrain by strip mining for coal. (U.S. Forest Service.)

Strip mining is rapidly becoming one of the most spectacularly visible forms of environmental alteration in the United States (Fig. 3.11). In Kentucky, West Virginia, and Pennsylvania, vast areas of the landscape are now being devastated by bulldozers in order to remove the layers of coal beneath. Various corporations, including the giant oil companies, are already involved and heavily invested in strip mining plans for future power production in these regions and in most of the other 23 states that have strippable coal deposits.

The Search for Underground Oil and Gas

Natural gas seems to have been found underground first in China over one thousand years ago. The search for salt by drilling led to the discovery of gas which was brought to the surface through bamboo pipes. In the United States, the search for underground salt deposits early in the 19th century resulted in extensive drilling operations that used a bit at the end of linked poles and wooden well casings. Such explorations sometimes encountered unexpected quantities of oil and gas.

When the first American well intended specifically for oil was drilled in 1859, petroleum did not have much practical value. But an early process of fractional distillation soon produced paraffin, kerosene, various greases, a gas, and a clear liquid, later to be known as gasoline. Of these the kerosene quickly became a principal source of lighting (kerosene lamps) in the United States, and the petroleum greases became vital lubricants, replacing vegetable or whale oil-based greases for train engines and other purposes. Gasoline was not to become important for several decades, until the invention of the internal combustion engine and the development of the automobile.

Fuels From Ancient Seas

Petroleum (crude oil) is a liquid of organic origin found trapped in pores and other spaces in certain sedimentary rocks in some regions. Petroleum varies slightly in composition from one region to another, but a typical crude oil may have about the following content: 82 percent of carbon, 15 percent of hydrogen, and three percent of oxygen and nitrogen taken together. The dominance of the first two elements leads to the term *hydrocarbon,* often used for petroleum as well as for natural gas, which is often found associated with crude oil.

The hydrocarbon molecules of crude oil and natural gas represent different combinations of the two elements. The distinguishing feature of each molecule of the various hydrocarbon substances is the number of carbon atoms per molecule. Natural gas or methane contains one carbon atom (to four of hydrogen); gasoline has between four and 12 carbon atoms per molecule; and fuel oils (for home heating) have between 15 and 18 carbon atoms per molecule.

It is generally agreed that petroleum and natural gas are of organic origin. All of the major occurrences have been found associated with sedimentary rocks of marine origin. The oil and gas appear to be the remains of living matter reduced by decay to a state in which carbon and hydrogen are the principal remaining elements.

Certain sedimentary rocks, such as dark-colored shales, contain high percentages of carbon of organic origin (some are called oil shales) and seem to represent the parent rocks for the actual accumulations of liquid petroleum or gas. These are concentrations usually found in more porous rocks into which the organic substances have gradually migrated, after moving out of parent rocks as droplets or molecules of hydrocarbon.

There is some uncertainty concerning the conditions under which the parent or source rocks of petroleum were laid down. A major hypothesis traces the hydrocarbons to microscopic-sized floating organisms living in vast numbers in ancient seas and preserved after death in accumulations at the bottom of these seas. Particularly suitable areas are bodies of marine water with only shallow connections with the ocean, having restricted circulation. Here tiny free-floating plants (algae) as well as floating or swimming animal life are abundant near the surface but, because of the enclosed nature of the water body, at greater depths the circulation is poor and oxygen depleted. This condition hinders oxidation and

Figure 3.12 Petroleum pumper brings oil to surface from a well drilled down to an oil reservoir or field in Wyoming. (Photo by author.)

decay so that organic remains that settle onto the bottom tend to accumulate along with the sediments.

The Black Sea is a large-scale example of a modern environment in which organic-rich sediment layers that may provide future oil and gas are now accumulating. Here the water circulates very slowly, and bottom sediments may contain a third or more of organic matter. Similar conditions occur in some of the Norwegian fjords, the Red Sea, and in the Persian Gulf.

On sea floors of better circulation but very rapid rates of sedimentation, organic matter may be buried more rapidly than it can decay and eventually become a source of oil. Parts of the Gulf of Mexico may be examples of such environments. Droplets of oil have been found below the surface in some Gulf sediments.

Draining the Fields

Migration: In a producing oil field, petroleum is recovered by drilling down to a *reservoir rock* (Fig. 3.12). This is usually a porous sandstone or limestone into which oil has slowly made its way up from the original source rock, often located far below and under pressure. Small amounts of oil seeping up from this source, year after year, have slowly accumulated in the overlying, porous rocks wherever they are surrounded or capped by some impervious rocks through which oil cannot migrate further (Fig. 3.13). Where such cap rocks did not prevent continued upward movement, the oil may reach the surface as a natural seep. But such natural petroleum seeps are so slow as to be worthless commercially.

It is the very slow, cumulative process of upward oil migration, proceeding through the centuries, and slowly filling the pores and cracks in overlying rocks, that has built up the underground reservoirs of the world's great oil fields.

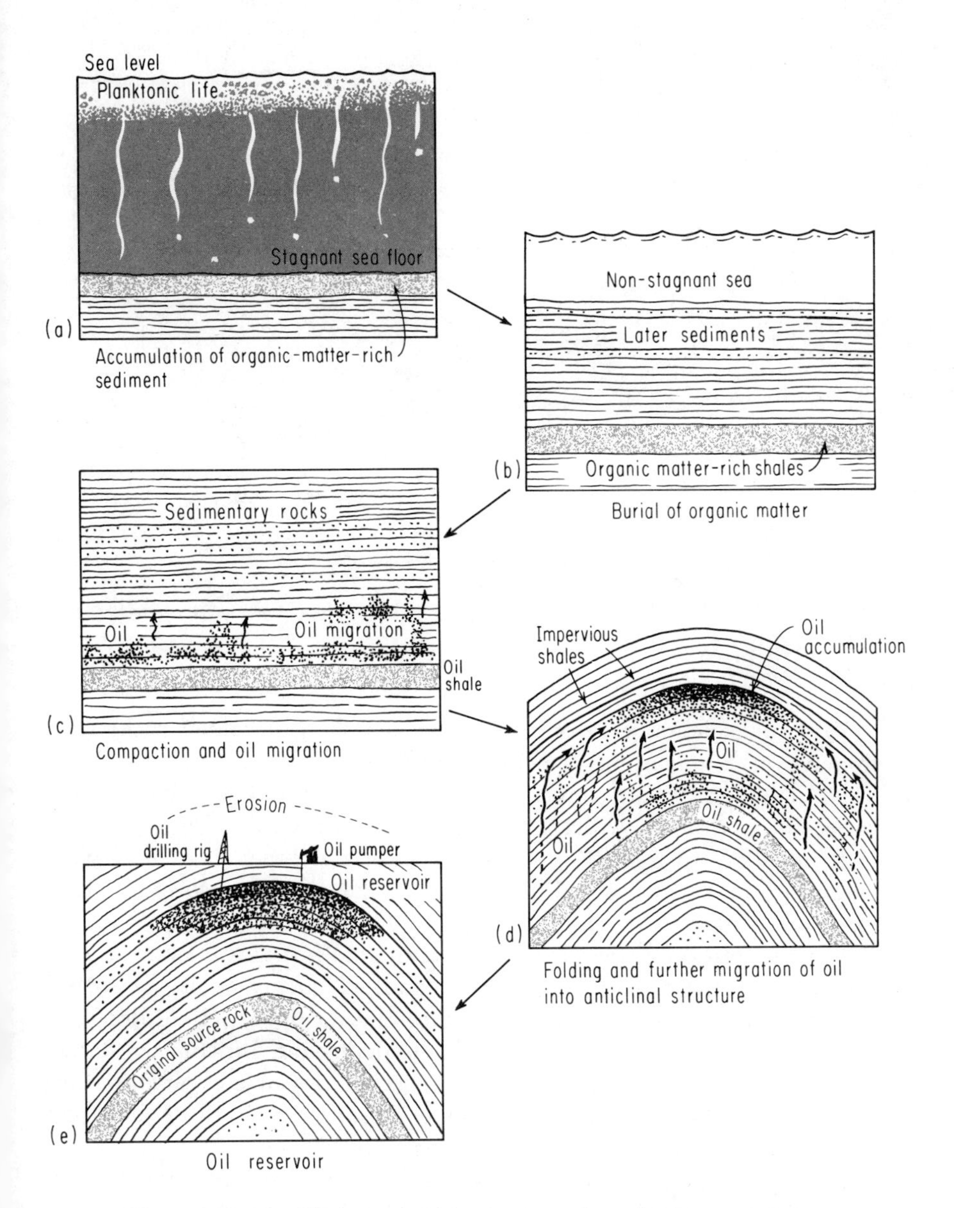

Figure 3.13 Simplified stages in development of a subsurface petroleum. (a) Accumulation of organic matter-rich sediment in suitable marine environment; (b) Burial of organic matter-rich sediment; (c) Compaction and early upward migration of petroleum substances; (d) Folding of layers and further upward migration of oil; (e) Collecting of oil in porous strata, capped by impervious layers forming oil reservoir.

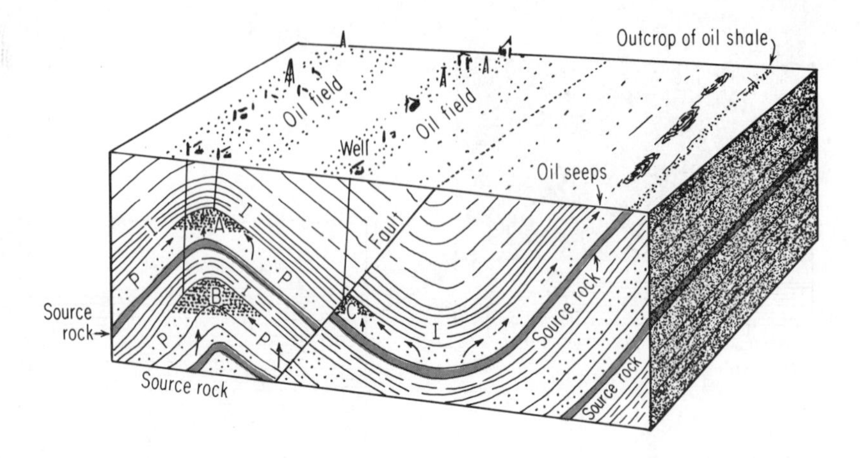

Figure 3.14 Oil reservoirs formed by folding and faulting of sedimentary rocks.

Trap: Petroleum seeping up through a region of folded rocks tends to move out of downfolds (synclines) and into upfolds (anticlines). Where an anticline or a dome consists of porous strata overlain by impermeable rocks, the petroleum will accumulate in the top of the fold, under the umbrella-like cap rock (Fig. 3.14). Methane and other gases may segregate from the petroleum and rise above the heavier oil to form a natural gas topping above the oil.

The recognition of the principle of anticlinal traps for oil was an early discovery of the petroleum industry and, for many years, petroleum geologists spent much of their time searching for oppositely dipping surface rocks that indicated the presence of underground anticlines or domes. Some deep-seated anticlines are not reflected by surface geology, however, and may be detected by complex exploration methods or sometimes encountered by random drilling ("wildcat wells").

Faults, or breaks along which rocks have been displaced, may truncate tilted, porous strata and bring impermeable rocks into such a position as to form a trap which becomes an oil reservoir.

World Distribution: Petroleum is produced on every continent except Antarctica. The major producing countries in the western hemisphere are Venezuela, United States, Canada, Mexico, Argentina, and Colombia. The major producers in the Middle East are Kuwait, Iran, Arabia, and

Iraq. In Africa, the production is from Libya and Algeria with new fields recently discovered in Nigeria. In Europe, Russia produces more oil than any other country. Indonesia produces most of the oil in the Pacific area.

In spite of the widespread distribution of petroleum, the greatest volume of proved reserves is concentrated in the Middle East. But new discoveries of oil are being made each year under the sea, on the continental shelf regions off several continents. Many geologists believe that more oil is to be found under these shallow waters than remains on all of our land areas.

Limits: Oil and gas fields are small compared with coal deposits. Before the beginning of man's use, the supply of coal was probably about 10 times that of the supply of petroleum liquids and natural gas combined.

The rate of accumulation of oil in reservoir rocks is very slow. In the time it takes for hundreds of barrels of oil to be pumped up from the average oil field reservoir, a few new *drops* may find their way up to the reservoir from far below. Indeed, it may be said that for all practical purposes our rate of extraction of oil is so much greater than the rate of accumulation, that petroleum, once used, is gone for all times. The same can be said for natural gas.

The world's demands for crude oil and natural gas have been almost doubling every ten years. If this continues, the world's petroleum and gas supply will be seriously depleted by the end of the present century. Using even the *most* optimistic estimates of world reserves, the world's supply of liquid petroleum will be gone in less than 100 years from now and natural gas long before that.

Mining Solid Oil

A major potential source of petroleum, known for years but now beginning to be used, consists of tar sands. These are sand deposits that have been saturated with heavy, large-molecule hydrocarbons. This asphaltic substance will not flow, remaining solidly filling the spaces between sand grains. In order to recover the tar, the rock is mined, heated with steam to allow the asphalt to flow and to be extracted. It must then be processed to recover the oil fraction of the substance.

The largest deposit of tar sands lies in northern Alberta along the Athabasca River, covering a region over 25,000 square miles. The Canadian tar sands have begun to be mined in the Fort McMurray vicinity and have yielded thousands of barrels of petroleum a year.

Another solid source of oil, not yet being used commercially, are the oil-shales that occur at or near the surface in various parts of the world, especially in North and South America. The largest known oil-shale deposits are the Green River Shales found in southwestern Wyoming, western Colorado, and eastern Utah.

The solid hydrocarbons in oil-shales may be extracted by a distillation process in which the rock is heated and the vapor obtained is cooled and condensed into liquid oil. Petroleum concentrations in oil-shales are highly variable as are the calculations of the ultimate production potential of these rocks. Most estimates of the petroleum that could be recovered from the Green River Shales range between 50 and 100 billion barrels.

At this time the oil-shales are not being used as a source of oil, and the actual price of producing oil from oil-shale is somewhat greater than that of producing from the tar sands or of producing crude oil from conventional wells. It seems likely that the bulk of our petroleum will continue to come from conventional sources for some time before the industry turns to oil-shale.

A major problem to be faced in mining these solid sources of oil will be the results of digging up and processing enormous quantities of rocks. Economic considerations will undoubtedly dictate strip mining procedures. The removal and piling up of overlying waste rocks and the disposal of the debris left after treatment of the oil-shales or tar sands, as well as the pollution aspects of dumping the waste, all promise serious environmental damage analogous to that now occurring from the strip mining of coal.

Natural gas is the cleanest and best of the fossil fuels from the standpoints of convenience and the minimal pollution side effects. Gas is generally produced from oil fields, though there are some fields that produce no gas and some gas fields that produce no oil. In the early days of petroleum production, it was not uncommon to allow natural gas to escape into the air when it was encountered. Laws now prohibit such waste in most states.

Gas can be moved through pipes more readily and rapidly than petroleum, and this is the usual manner of its transportation from the producing to the consuming regions. For this reason, using natural gas instead of petroleum eliminates the need for air-polluting fuel trucks and the danger of oil spills from tankers on the seas.

It was not until after World War II that large-diameter pipelines allowed effective long-distant transmission of natural gas from Texas and Gulf Coast regions to the major consumer areas of the north and northeast. In the last 25 years, natural gas has become so popular a fuel that there are now growing signs of depletion and shortages. As we project future demand, it appears that effective exhaustion of the major American sources of natural gas may occur within a decade or so. Ships fitted with large pressure tanks will soon be bringing vast quantities of natural gas in compressed form from North Africa to the United States to avert the impending shortages of domestic gas.

Prior to the arrival of natural gas two decades or so ago, large cities like New York used coal gas for cooking and other purposes. Many parts of Europe still depend on this gas. Coal gas is made by heating coal with a limited quantity of air. Unlike natural gas, it contains nitrogen and other unburnable substances so that burning a cubic foot of coal gas liberates less heat when burned than does the same quantity of natural gas.

It now appears that gas produced from coal will become an increasingly important fuel for domestic and industrial purposes. It will at least partially replace natural gas that is far less abundant than petroleum or coal, even when taking into account foreign reserves like those of North Africa.

NUCLEAR ENERGY

The Nuclear Era

Atomic fission, the artificial disruption of the nucleus of an atom, was first achieved in 1939 in Germany when the isotope uranium 235 was split by bombardment with a neutron. The result was the production of two lighter elements, barium and krypton, as well as three new neutrons. In 1942, Enrico Fermi was able to achieve a fission procedure in which the number of uranium atoms was great enough (critical-mass) that the neutrons freed by fission were numerous enough to act continuously on the remaining uranium to produce a *chain reaction* (Fig. 3.15). The fission of uranium 235 was the basis for the first atomic bombs and has been controlled in modern nuclear reactors that utilize the immense amount of heat released by the fission of uranium to form steam, turn generators, and produce electricity (Fig. 3.5). Uranium is an immensely concentrated source of energy. The energy released by fission of one pound of uranium is as great as that produced by burning several thousand tons of coal.

The first electric power from atomic fission was produced in 1951, and the first major nuclear-electric installation went into operation in 1957, in Pennsylvania. About two dozen nuclear power plants were operating late in 1971, with 50 more under construction (Fig. 3.16). It has been predicted that, by the year 2000, 50 percent of the energy produced in the United States will be from nuclear fission.

Uranium is the heaviest of the naturally-occurring chemical elements.

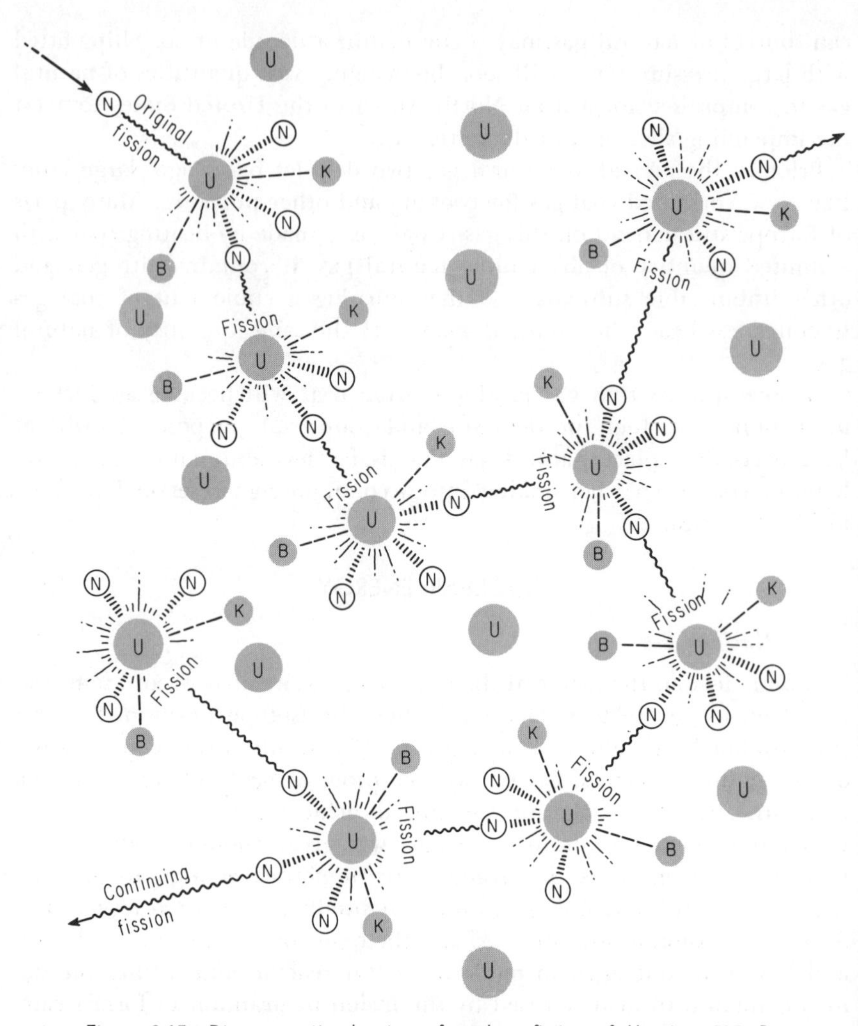

Figure 3.15 Diagrammatic drawing of nuclear fission of Uranium 235. B and K represent fission (daughter) products; N represents neutrons freed by fission that continue the reaction by encounter with other uranium atoms (chain reaction). Heat, which is also a by-product, is used to produce steam in nuclear power plants and to generate electricity.

As it occurs in rocks, uranium consists of a mixture of three isotopes: uranium 238, uranium 235, and uranium 234. The amount of uranium 234 is so tiny that it may be ignored. For practical purposes, natural uranium may be considered as consisting of atoms of uranium 238 and uranium 235 in the ratio of one atom of U235 to 140 atoms of U238. In other words, more than 99.2 percent of uranium consists of the isotope uranium 238; much less than one percent is uranium 235. But uranium

Figure 3.16 Nuclear-fission power plant (burner) at Indian Point, New York. Electricity produced is used in New York City. The Hudson River receives the hot waters discharged. (Photo courtesy of Consolidated Edison of New York.)

235 is unique in that it is the only naturally occurring isotope that is spontaneously fissionable by the capture of slow neutrons. For this reason, uranium 235 is the initial fuel of all fission reactions.

Fission Reactors

If nuclear energy depended entirely on the uranium 235 isotope, the nuclear fuel interval would be brief, because of the scarcity of this form of uranium. However, a few reactors have been developed in which uranium 238 is transformed into the fissionable isotope plutonium 239, thereby creating nuclear fuel as they operate. Such atomic reactors are called *breeders;* ordinary uranium 235-using reactors can be called *burners.*

In the burner reactor all the fissionable material, the uranium 235 is used up as the fuel consumption cycle is completed; the uranium 238 is wasted. In a breeder reactor, some of the uranium 238 atoms, which form over 99 percent of all uranium, absorb neutrons and are converted to uranium 239. This new isotope, by two rapid radioactive transformations, breaks down into plutonium 239. Plutonium 239 is fissionable in a manner similar to uranium 235.

The ability of breeder reactors to utilize uranium 238 opens the prospect of making a fuller use of the world's uranium ore reserves than if only the scarce 235 isotope were used as occurs in common burner reactors. Unfortunately, all of the commercial atomic power plants in operation in the United States in 1971, and most of those under construction were burner reactors!

Finding Uranium

The measured resources of rich uranium deposits are unfortunately not as large as was expected after the "uranium rush" of the 1950s. High-grade deposits are relatively rare in the United States and are largely concentrated in Mesozoic sedimentary rocks of the Colorado Plateau in Utah, Arizona, Colorado, New Mexico, and in Cenozoic rocks of Wyoming. These are deposits rich enough to be used readily in the burner reactors. In the entire world, there are probably less than one million tons of such rich ores.

If the burner type reactors continue to dominate among atomic power plants and the use of nuclear energy grows at the rapid rate predicted, there will be such a heavy drain on the supplies of uranium that high grade ores in the United States may be exhausted by about 1990. But if burner reactors are phased out and breeder reactors rapidly brought into large-scale usage, much lower grade, more abundant uranium deposits can be used because most of natural uranium (the 238 isotope) can be used in the breeders and there need not be as high a uranium percentage in the rock being mined.

Low-grade uranium ores, those that can be used in breeder reactors, are relatively abundant. A major example is the Chattanooga black shale deposit that occurs along the western edge of the Appalachian Mountains in eastern Tennessee and adjacent states. This large low-grade deposit by itself could theoretically supply fuel equivalent in energy to all of the fossil fuels found in the United States. This is quite hypothetical because in order to use all of the Chattanooga shale, it would be necessary to strip mine hundreds of square miles of the Appalachian landscape—a dismal prospect, unlikely to appeal to inhabitants of a region already being devastated by strip mining for coal.

In order to derive most of our future energy from nuclear fission, it is absolutely essential that successful breeder reactors be developed soon, probably within the next decade or two. Unfortunately, the present American nuclear reactor program consists of "burners" that will rapidly exhaust the reserves of uranium 235-rich rocks. There is no certainty that breeder reactors will be developed in time to prevent the essential exhaustion of all high-grade uranium deposits.

The Hope of Fusion

In recent years scientists have been trying to perfect atomic *fusion,* a reaction in which simple atoms are converted into more complex, heavier atoms by the joining or fusion of the nuclei of the lighter atoms. The source of the energy of the sun and other stars is the continuous fusing together of hydrogen atoms into atoms of helium 4. A similar reaction has been accomplished by physicists who combined *deuterium,* an isotope of hydrogen with a mass of 2 with *tritium,* an isotope of hydrogen with a mass of 3, to form helium, under high temperatures. Unfortunately, this is an explosive reaction, that of the hydrogen bomb! Research to achieve *controlled* fusion has been going on for two decades. Some progress has been achieved, but it is still not possible to estimate when or even whether the development of power from the fusion reaction may ever become a reality.

The most feasible possibility for controlled fusion seems to be the deuterium-tritium reaction, rather than a deuterium-deuterium reaction, another fusion possibility, although the latter would be more desirable because of the great abundance of deuterium (one atom for every 6700 atoms of hydrogen). By contrast, tritium is extremely rare in nature so that any considerable amount must be manufactured for lithium 6, a minor component of natural lithium. Lithium may be obtained from some granites and other igneous rocks as well as from some salt deposits. The lithium 6 isotope must be considered a rare substance. Its total supply is uncertain but definitely tiny in comparison with deuterium.

It is impressive to speculate that were controlled fusion of the deuterium-deuterium variety to become an accomplished fact, the energy obtainable from the deuterium in just 20 cubic miles of sea-water would be about equal to that of the earth's total original supply of all fossil fuels. But judging from present research trends, it does not seem likely that we will use atomic fusion as a major source of energy for many decades, if at all.

ENERGY AND POLLUTION

Pollution and other forms of environmental damage follow inevitably from the production of vast amounts of energy by burning fossil fuels or by atomic fission (Figs. 3.17, 3.18). As the demand for energy soars in a region, power plants become larger, more numerous, and ever more conspicuous features of the American landscape. Just the space occupied by modern power plants is becoming a problem in some regions. A million kilowatt nuclear generating plant, for example, requires between 500 and 1000 acres of land. The high voltage transmission lines that

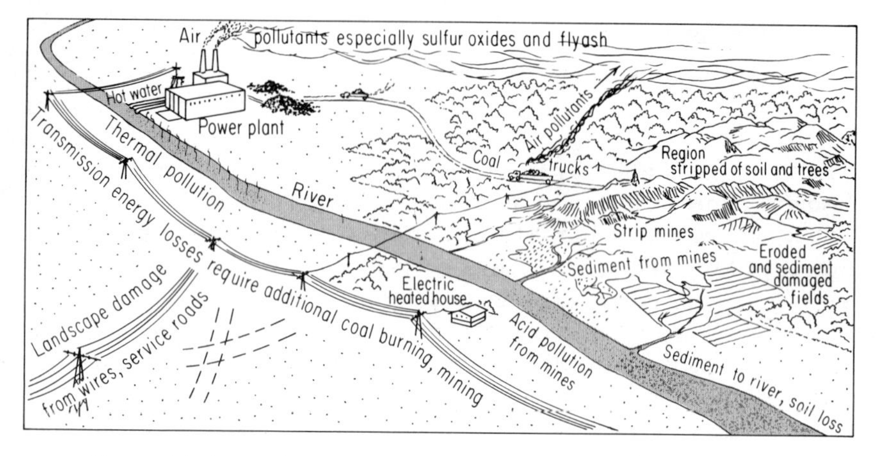

Figure 3.17 Various kinds of environmental damage caused by generation of electricity include air and water pollution and landscape deterioration from strip mining of coal.

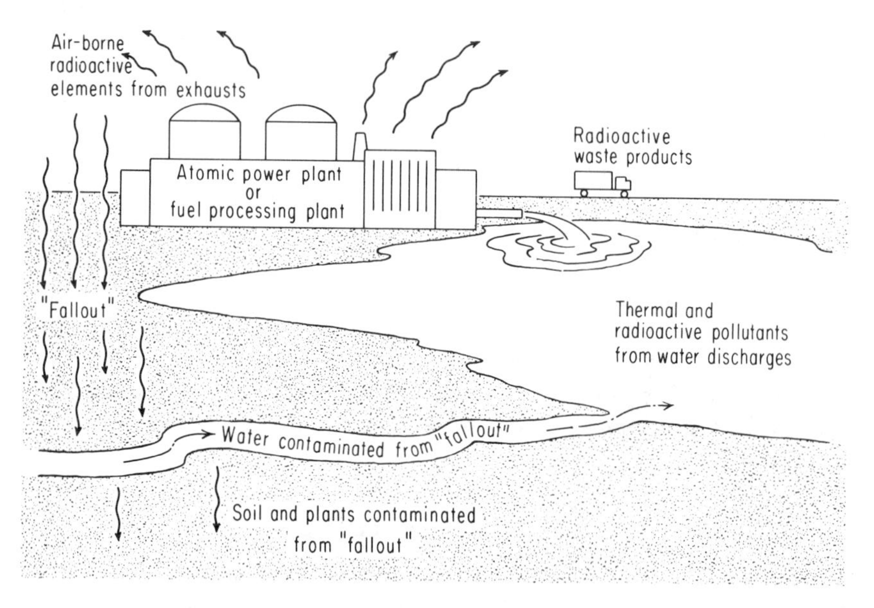

Figure 3.18 Various kinds of environmental hazards caused by generation of electricity by nuclear fission.

radiate out in various directions from such a plant require rights of way 750 feet wide across the landscape, thereby removing about 70 acres for every mile of transmission line from other productive uses. The amount of useful space absorbed by transmission lines stretching hundreds of miles is enormous. The scenic damage done to a landscape by the great scars left by transmission lines is incalculable (Fig. 3.19).

Figure 3.19 Electrical transmission lines are a major cause of landscape damage in the United States. Wide swath and felled trees mark the course of these power lines cutting across National Forest land in Colorado. (Photo by author.)

Up the Stacks

Haze and Gases: A large part of the air pollution in the United States arises from our high per capita use of energy. The United States is covered by a hazy veil of suspended particles visible almost anywhere from a high vantage point and under the proper lighting conditions (Fig. 1.1). In and near urban and industrial regions, the concentrations of floating particles and of gases such as sulfur dioxide, carbon monoxide, carbon dioxide, and nitrous oxides are especially heavy, sometimes enough so to be fatal, especially to the aged or those with pulmonary problems (see Chap. 1). As the demand for energy soars, the fossil-fuel power plants step up production and proliferate. The result is an ever-rising contribution by the power-producing companies to the nation's air pollution (Fig. 3.3).

Many of the pollutants that arise from the burning of fossil fuels, such as sulfur dioxide, tend to concentrate in the region where they are generated and may have a relatively short life. But carbon dioxide is in another category. The combustion of fossil fuels, no matter how efficiently done, always produces carbon dioxide. Its concentration in the atmosphere has been increasing in recent years by about 0.25 percent a year.

A number of scientists are becoming worried about the ever-increasing level of carbon dioxide because of its effect on the world's climate. This problem was considered in Chapter 1 under "Weather and Climate Alteration."

The tendency is for increasing carbon dioxide levels to cause a rise in world temperatures but for the increasing particulate haze to cause a decrease in temperature. The degree to which these two tendencies may balance one another or not remains to be seen, and the subject is controversial at this time.

Chapter 1 includes a survey of the complex nature and multiple causes of air pollution.

New Techniques: A new means of generating electricity from fossil fuels, which is essentially nonpolluting, is the use of the magnetohydrodynamic (MHD) generator. Unlike existing steam generating plants, the new device produces power by burning fossil fuels at high temperatures. An electrically-conducting gas or liquid is sent at high velocity through a magnetic field. This procedure generates an electrical current without the need for steam boilers or conventional generators. As much as half of the energy value of the fuel used is transformed into electricity, as compared with steam power plants that may *lose* over 65 percent of the energy value of their fossil fuel during their generating process.

The world's first MHD generator went into operation in the USSR in 1972. MHD development in this country is proceeding slowly because of a general lack of official interest and a lack of funds. This revolutionary method of electric production seems capable of greatly reducing both the air pollution and thermal water pollution (see below) that have long been undesirable by-products of fossil fuel combustion.

Hotter and Hotter Waters

Damaging Our Rivers: The discharge of waste heat into natural waters is called thermal pollution, and the chief source of this waste heat is the electric power industry. Electric power plants are generally situated near rivers or other major water bodies because of the water required for steam production and in order to have a means of disposal of the hot water into which the steam condenses.

Fossil-fueled electric generating plants may discharge thousands of gallons of hot water *a second.* Nuclear power plants produce even larger quantities of hot water (per kilowatt generated) than fossil-fuel power plants because they operate at lower efficiency due to the lower temperature of their boilers. This means that some 70 percent of the energy used in a nuclear plant becomes waste heat. This usually takes the form of vast quantities of hot water poured into an adjacent water body at temperatures from 10° to 30°F higher than its original temperature.

The effect of discharging waste heat into a body of water containing animal life is not always obvious. But we know that some fish and other forms of aquatic life are killed by a moderate temperature rise, and others are rendered unable to reproduce or chances of their young surviving after spawning are reduced. Some aquatic organisms become more susceptible to disease as temperature rises.

Rapid changes in temperature are produced by the variation in waste heat discharges of many power plants. These changes may be even more dangerous to the metabolism of aquatic life than the heat alone. Disruption in living habits and migration patterns frequently results from heat discharged into rivers. Plant growth in shallow water may be stimulated by heat to the point where it restricts water circulation, increases silt accumulation, and interferes with animal life. The ability of a river or lake to break down and purify sewage and other wastes can also be considerably reduced by a rise in temperature.

It is becoming ever clearer that because of the recent growth in electric power generation, many of our inland water bodies are approaching a natural limitation in their ability to absorb waste heat without damage to aquatic life.

Regulation and Solutions: In recent years, the recognition of the various dangers of thermal pollution has led to the development of regulations against pouring power plant discharges into bodies of limited circulation such as Lake Michigan and Biscayne Bay, Florida.

A number of power plants are now under construction along the seacoast, and future power plants may be located several miles offshore so that waste heat can be dumped into the surrounding ocean and dissipated more readily than in rivers.

In order to solve the thermal pollution problems of power plants that have to be located inland, waste heat may be reduced by the use of *cooling towers* (Fig. 3.20). One type of cooling tower employs the principle of removing heat by evaporation: the heated water is discharged into a tower some 300 to 450 feet high where it falls as a thin film over a series of baffles exposed to air rising through the tower. In another process, the heated water is sprayed into the tower as a mist that evaporates and cools quickly. In both cases, the water (minus the fraction lost by evaporation) is collected in a basin, at a temperature lowered by about 20°F and then discharged into a nearby water body.

A major drawback of cooling towers is the large amount of water vapor discharged into the atmosphere during the cooling process. A tower for a 1000 megawatt power plant would lose over 20,000 gallons

Figure 3.20 Cooling towers are an effective means of decreasing thermal-water pollution from power plants. (Photo by author.)

of evaporated water per minute. On cold days, such a discharge can condense into a thick fog and cause icing-over in the area in which the plant is located.

By 1970 all steam-electric plants in England were required to have cooling towers. Yet, most American power companies avoided the use of cooling towers on the grounds of high cost, the space required for such structures, and the possibilities of adverse effects on the atmosphere.

Useful Purposes?: The waters used by power plants for cooling purposes already comprise about 10 percent of the total water flowing over the American landscape. In two or three decades, we may be heating up as much as $\frac{1}{3}$ of the nation's fresh waters as the result of the operation of electric power plants. Obviously, thermal pollution on such a scale is not feasible and new solutions must be sought.

If heated power plant waters could be used for some useful commercial or domestic purposes, then the fuel that would have been used for energy for these purposes would not have to be expended, thereby reducing overall power plant use and decreasing pollution.

Discharge waters from power plants are usually not hot enough to heat buildings, but some plants near industrial centers do sell the low pressure steam escaping from their electric-producing turbines, before it condenses to hot waste water, to chemical refineries and industrial plants. In New York City, such steam is used for heating thousands of buildings including the skyscraper office structures.

Heated water from power plants may someday be used to accelerate the growth of certain fish and shellfish, or even to provide ice-free shipping lanes, in some regions. But, at least for the present, there seems to be no large-scale, practical use for the hot waters that are pouring in ever-larger torrents from ever-increasing numbers of power plants across the nation.

Brave New Radioactive World

Threats of Nuclear Power: Electric utility companies have proposed building 700 nuclear power plants, of at least one million kilowatts each, by 1980. Over 50 nuclear reactors were being built in 1971 to be added to the two dozen already in operation (Fig. 3.16). Unfortunately, there are many unanswered questions regarding the environmental impact of the present reactors and concerning the possible dangers to man.

We should not disregard the possibility of explosions or serious accidents in atomic power plants. Indeed, the risk rises as the number of reactors in operation increases and is greater for fast breeder reactors that use liquid sodium for cooling and contain vast quantities of deadly plutonium 239. The number of fast breeders must increase proportionately to the somewhat safer burner reactors, for reasons of their ability to utilize the relatively abundant U238 isotope rather than the rarer U235, as noted earlier in this chapter.

However, most of the questions of nuclear safety that are now being asked by scientists and environmentalists are of a nature more subtle but more probable and immediate than the concern with possibilities of accidental explosions in reactors. These are questions of radioactive pollution from *normal* operation of nuclear reactors (Fig. 3.18). Such pollution may take two principal forms: (1) reactor residue wastes, "leftovers" of nuclear fission; (2) low-concentration radioactive isotopes which escape into air and water during reactor operation.

Deadly End Wastes: Radioactive waste products are produced by all nuclear fission reactors. These must be removed periodically and kept from contact with all life forms. Hazards exist in two forms: first, there is the process of transportation itself, involving the danger of accidental contamination along the route; second, there is the question of location of the final storage sites for these mixtures of highly radioactive isotopes. Future generations will have to watch over hundreds of millions, probably billions of gallons of radioactive wastes seething away in underground storage, some wastes so hot they will boil for hundreds of years.

One possibility for relatively safe, permanent storage of wastes is to change the liquid isotope mixtures into solids in the form of glass or ceramic masses that can then be buried underground. In general, solidification and underground storage must be considered a preferable alternative to the past practice of ocean dumping. But there is considerable controversy about the safety of underground storage of nuclear wastes. In some regions, there is risk of earthquake damage, and in almost any region, there exists the possibility of contamination of local ground water and the surface rivers toward which they flow (see Chap. 2). The Atomic Energy Commission has relied in recent years on storage in abandoned salt mines, salt being dense and relatively impervious to water. However, many geologists dispute the safety of this procedure. By 1972, large amounts of underground water had been discovered in abandoned mines very near to salt mines already being used for storage by the A.E.C. Even with better selection of salt mines or other sites, it is doubtful that any underground storage area can be guaranteed safe from water seepage for the enormous lengths of time during which the isotopes remain radioactive.

Many scientists now feel that the dangers posed by the great quantities of radioactive wastes to be produced by future nuclear fission development are so great that storage *anywhere* in our environment is too great a risk to our survival. Serious proposals have been made recently that concentrated radioactive wastes be disposed of outside of the planet. The "space shuttle" project being developed in 1972 might make it possible one day to dispose of the deadly substances outside of the planet by transporting it to man-made satellites and, thence, on to its extraterrestrial destination, perhaps in the sun itself.

Radioactivity Around Us: Nuclear power plants routinely release controlled, small quantities of radioactive gases or particles to the atmosphere and radioactive liquids to natural bodies of water. In the United States, such emissions are generally kept below concentrations set as safe by Atomic Energy Commission policies. These limits are not known for certain to be absolutely safe for humans and are not at all designed for the protection of aquatic life or any organism other than man.

The radioactive isotopes released into air and water by nuclear plants are usually in very low concentrations. *Dilution* in the environment is the key to the hope that such substances do not constitute a serious hazard. Unfortunately, there are several flaws in the reasoning that the discharge of minute quantities of radioactive substances is not detrimental to life. First, the small quantities released from a single nuclear plant must be multiplied by hundreds, eventually thousands of times, as new plants proliferate throughout the country. The longer-lived isotopes settle down and accumulate in the soils, build up in lakes, and may even concentrate high in the atmosphere, far beyond the levels considered safe in the discharges of a single plant (Fig. 3.18).

Another very important consideration in evaluating the real dangers of radioactive operating discharges is the tendency of organisms themselves to concentrate isotopes internally to higher levels than exist around them. For example, the eggs of ducks living on the Columbia River downstream from the Hanford nuclear power plant in Washington showed a concentration of radioactive phosphorous 32 that was 200,000 times greater than that of the river itself. Rabbits in this vicinity had concentrations of radioactive iodine 131 in their thyroids at a level 500 times that found on the vegetation around the plant.

The various isotopes produced by nuclear reactors tend to accumulate in different organs or parts of the body just as iodine 131 is stored (along with stable iodine) in the thyroid glands. Strontium 90 is picked up from the environment by plants and animals and used as the equivalent of natural calcium. It becomes part of bones and some tissues, damaging the growing cells of the bone marrow, and may give rise to leukemia or other forms of cancer. It is sometimes found heavily concentrated in human mother's milk. Cesium 137 is another radioactive by-product of nuclear reactors that tends to build up in the body, especially in the muscles and other soft tissues.

The length of time that a radioactive element persists is suggested by its half-life, the time required for half the atoms in any given quantity of radioactive material to disappear by change into other elements. Iodine 131 has a half-life of only 8 days. But its newly-studied cousin, iodine 129, also a by-product of nuclear power, has a half-life of 17 million years! Strontium 90 and cesium 137 have half-lives of 28 and 33 years, respectively, so that even though rates at which these last three isotopes are produced may be very low, they will build up over the years and may eventually reach dangerous levels.

Uranium mining presents another radioactive environmental hazard. The danger was first noted in the late 1930s with the early deaths of many of the pitchblende miners of Germany and Czechoslovakia. Pitchblende, an ore of uranium, is also a source of radium, for which it was then being mined. In this region 90 percent of the miners eventually died of lung diseases, chiefly cancer, apparently from the inhalation of radon gas that is a product of the natural radioactive decay of uranium. Studies have also shown a very high incidence of lung disease among American uranium miners of the 1950s and 1960s.

Another danger of uranium mining comes from the piles of waste rock or tailings from the mines and processing mills. These tailings contain highly toxic radium 226 that is readily leached from the loose rock by rainwater and enters streams in solution. It has a long half-life, remaining hazardous for thousands of years, and decays to radon gas that can concentrate in buildings constructed on tailings used as landfill or used in concrete walls. In 1971, about a thousand buildings in Grand Junction, Colorado were discovered to have been constructed on radioactive tailings taken from conveniently nearby uranium workings. Many showed high levels of radon gas.

A Radioactive Future: Various ordinary substances are radioactive in some degree, and all life on earth is subjected continuously to a low level of radiation from naturally radioactive elements in their immediately surrounding environment and from cosmic rays. This is "background radiation," which indeed may have been a factor in the production of the genetic mutations that were part of the gradual evolution of life and of man.

There has been much scientific speculation concerning the role of natural radiation in the aging process, in cancer and in other physiological deterioration processes. On a day-to-day basis, the body is able to function and cope with the low degree of tissue damage produced by background radiation. But if the radiation rate is significantly increased, it may reach the levels of permanent injury.

There is no way of knowing what the cumulative effects on humans will be of the continuous intake of minute quantities of various radioactive isotopes from our water, air, and food over decades of life. Unlike most other environmental hazards, the danger from radioactive isotopes concentrating within the body is new and unclear. For example, 30 years ago there probably was no strontium 90 on earth. Today all living creatures, from pole to pole, contain this dangerous substance in their bodies. What will happen when the infants of the 1950s reach the end of this century after a lifetime of slowly increasing and cumulative doses of half a dozen hazardous isotopes? And, considering the genetic effects of radiation, what will their descendants be like?

ALTERNATE ENERGY SOURCES

Great expectations have been held out for nuclear power as an ultimate solution to the problem of air pollution from fossil fuel combustion and as a replacement for fossil fuels when they have been exhausted. But in recent years, there has been growing awareness of the environmental dangers of nuclear waste products and of the inability of science and technology to achieve a fusion reaction not dependent on scarce uranium.

While many scientists continue to try to achieve the technological means of producing "clean and unlimited" nuclear power, others have been re-examining natural processes in a new search for nonpolluting and even nonexhaustible energy. We will consider next the nature and practicability of four such sources of power: the winds, the tides, geothermal energy, and direct solar energy.

Windpower

Windmills were first developed in Western Europe in the 12th century for grinding grains and pumping water. The great sails of windmills have long been familiar to travelers in Holland and other flat and windswept regions of Europe. In the United States, a smaller, more efficient type of windmill was vital to the cattle and farms of the Great Plains (Fig. 3.21). Tens of thousands of these were built in the period from 1873 to about World War I for the purpose of pumping water from far below the dry prairie land. Almost all have been abandoned since electrification has occurred.

The two great disadvantages of windpower is the undependability of wind velocity and the difficulty of storing the energy derived during the times of strongest winds. These are unfortunate problems because the energy of the wind is free of cost; once the mill has been constructed it

Figure 3.21 Wind power is pollution free. This windmill in Utah pumps water for cattle in a region remote from electric-power lines, but it pumps only when the wind blows. (USDA Soil Conservation Service.)

can be used without causing waste products or pollution. The potential of this energy source is considerable, especially in windy regions where wind energy may average over 20 watts per square foot. Wind-driven electric generators have achieved efficiencies as high as 75 percent.

One proposal for storage of wind-derived power is to use the output of a wind-driven generator to decompose water into hydrogen and oxygen. These would be compressed, stored, and recombined in a fuel cell, as needed, to generate a steady flow of electricity, much as a reservoir can provide a steady supply of drinking water for a region in which the rainfall and stream flow are very intermittent. The hydrogen could also be burned in a gas turbine, thereby turning a conventional generator of electricity.

This scheme for wind use is still far from realization, but there is no question that modern wind-driven electric generators could now be used to *supplement* conventional fossil-fuel power plants in order to reduce the consumption of coal or oil or to boost power output during times when peak demand coincided with relatively steady winds. But few energy planners are likely to consider this combination as dependable enough to include in new construction. Nor is it likely that much attention will be given to wind-driven generators using the decomposition of water as a means of storage, as long as our energy demands continue to rise at the rates we have seen in recent years. However free of pollution and low of cost, wind power seems a relatively minor source of energy for the modern world.

The Eternal Tides

Twice a Day: The possibility of harnessing the tides to do useful work has long fascinated dreamers and scientists alike. More dependable than the winds, the tides are eternally rising and falling. Twice every day the surface of the sea goes from high to low level with about 12½ hours between each tide. These movements are the result of the gravitational attraction of the moon (to a lesser extent, the sun) on the ocean waters as the earth rotates.

The alternate rise and fall of sea level produces long, slow waves that spread out in all directions from the equator. These tidal bulges produce no noticeable effect in the middle of the ocean where sea level changes are only a few feet. But along coasts where there are bays and other indentations, the tides may pile up water to considerable heights.

The shoreward-moving tidal current is called the flood tide. When these currents flow into a funnel-shaped arm of the sea such as the English Channel, the result is a greater rise and fall of the sea level than in the open sea. Even higher tides occur when these currents run up into a narrowing channel. In Canada's Bay of Fundy, high tide is usually over 50 feet higher than low tide, meaning a change of some 50 feet every six hours.

Ocean Dams: Tidal power is obtained by building dams across the mouth of bays or tidal estuaries. Small dams powering mills used for the grinding of grain have been in use since about the 12th century.

A modern tidal power plant, in simplest form, consists of a low dam with sluice gates across the mouth of a tidal basin. The rising tide is allowed to flow through the open gates into the basin (Fig. 3.22). At high tide the gates are closed and, as soon as the sea level has dropped enough to create a usable head of water in the basin (over 10 feet), the trapped water is gradually released through the generators in a power plant built next to the dam. This simple type of tidal power device has the disadvantage of an intermittent input of energy.

More dependable power output is achieved by using turbines designed to operate in either direction and using the inflow through the sluice gates as the basin is filling as well as the outflow after it is full. Another approach to the problem is that of the twin or multiple basin system that provides a *constant* working head between the two or more separate basins and, thus, makes the power output relatively independent of the times of high and low tides.

Only within recent decades have tidal-electric installations been given serious engineering consideration and not until very recently have any actually been brought into operation. One of the best known tidal power projects is located on Passamaquoddy Bay on the United States-Canadian boundary, off the Bay of Fundy. Passamaquoddy Bay is an irregular inlet that lends itself to a multiple basin type of power station. The power project here was first proposed in 1920 with some construction occurring in the 1930s; it was resurrected in 1963 but held up from completion by Congressional failure to vote the necessary funds.

France has succeeded in developing the world's first commercial tidal power station that opened in 1966 on the Rance Estuary. This power plant has a capacity of 240 megawatts, and the French are working on plans for a much larger project in the Mont-Saint-Michel Bay not very far away from the Rance.

Limits: Unfortunately is is not possible to harness tidal energy everywhere. In practice, any power we can develop from tidal action proves uneconomic wherever the tidal range averages much less than 25 feet or in places where the coastline lacks long, narrow, enclosed bays or estuaries that can be readily dammed. Sites fulfilling these conditions are to be found along only about *five percent* of the world's coastline.

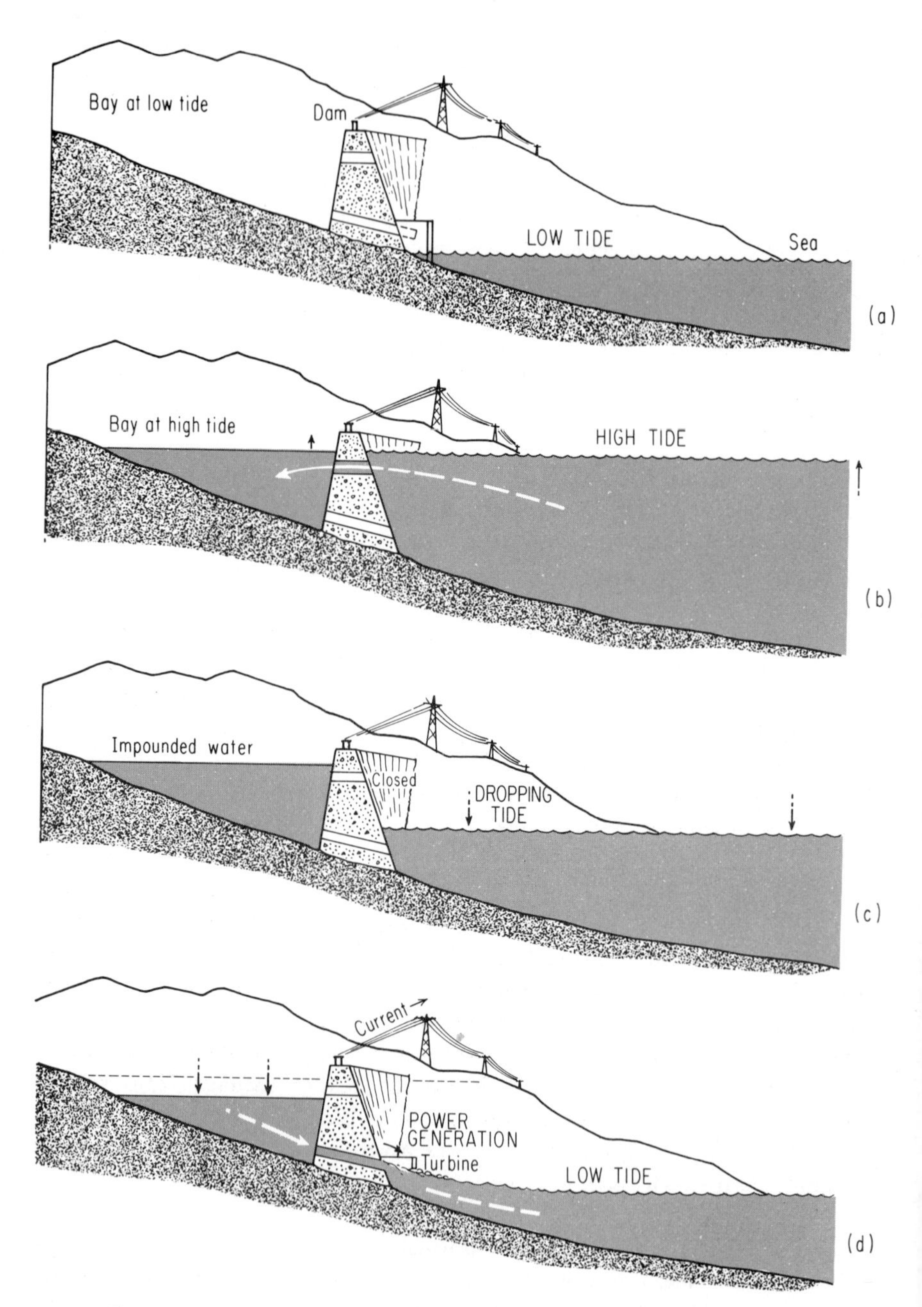

Figure 3.22 Tidal power is pollution free but can only be harnessed where seacoast irregularities and tidal differences permit. Simplified diagram shows principle of tidal dams. Water fills dammed bay at high tide; dam is closed as tide drops, and flow of dammed water through turbine generates electricity.

Considering these several factors, it has been estimated that the world's potential tidal power, if fully developed, would amount only to about one percent of the potential of the world's running water power, were *it* fully developed. This is a very small fraction of modern energy demands. Nevertheless, tidal power is capable of being developed in certain favorable localities to the extent of hundreds, possibly over a thousand, megawatts per site. Most important, tidal power has the advantage of producing no noxious wastes, consuming no exhaustible fuels, and creating much less disturbance to the landscape than the dammed reservoirs required by hydropower installations. From the standpoint of environmental benefits, all possible tidal power projects should be carried to completion with high priority.

Heat from Below

The earth's temperature gradient in the upper part of the earth's crust is about 1°F rise per 100 feet of depth. In the deepest wells, which reach depths of over two miles, temperatures of several hundred degrees F have been encountered, but in certain volcanic regions underground water concentrated in porous or fractured rocks only a few hundred feet below the surface becomes superheated from contact with hot, volcanic rock. Wells drilled into such concentrations of hot water will conduct water at above boiling point upward toward the surface. As the water rises, it changes into steam at high temperatures and can be used for heating buildings. More important, it can be used as the energy source for a conventional steam-electric power plant, bypassing the need for burning fossil fuels.

The earliest utilization of this *geothermal* energy was at Larderello, Italy in 1904. The capacity of power plants in this region is over 400 megawatts. In New Zealand, a geothermal steam power plant that began operation in 1958 has a capacity near 300 megawatts. In northern California, a smaller geothermal power plant began operation in 1960 with considerable future expansion planned. Mexico has a very small plant north of Mexico City and has explored a region in Baja, California that may have one of the largest steam production potentials in the world. Iceland, a volcanic island that has long been heating buildings with water of geothermal origin, has only recently begun the production of geothermal electric power.

It has been calculated that the total potential for geothermal power is about the same as that of the world's potential tidal power, if both were fully developed at all suitable locations. However, unlike the inexhaustible supply of tidal power, geothermal energy obtained from volcanic and thermal regions depends on the heat which has been building up in these regions over many years. In other words, it draws from a resource that is only renewable on a minor scale. In practical terms, geothermal power, as understood and used today, is not only small in quantity when compared with recent energy demands but has a short-life expectancy. Most of this stored energy would be exhausted in half a century of heavy use.

The Solar Furnace

Most of our power today is derived from the fossil fuels, the stored remains of ancient life that owed their existence to the sunlight of hundreds of millions of years ago. So it is appropriate in our search for energy sources less exhaustible and less environment-damaging than the fossil fuels, that we turn directly to the sun, the fundamental source of energy in our solar system.

Collecting Sunlight: The total energy of the solar radiation reaching the earth's surface is enormous. In deserts and other areas of high annual sunshine, the daily solar energy input in one square centimeter of horizontal surface may average 500 calories or more. By using photovoltaic cells, it should be possible to convert about 10 percent of the solar energy into electrical energy. This means that to run a 1000 megawatt electrical power plant (the size of a large fossil fuel or nuclear plant) 10,000 megawatts of solar power would have to be collected. This would require a horizontal collection area of about five miles by five miles, if square in shape. For regions of lower average sunshine, the size would be proportionately higher.

The solar energy reaching the arid regions of the west is so great that if a region about a quarter of the size of Nevada were set aside for the collection of solar energy, it should be possible to generate enough power to supply the nation's power demands until the year 2000. Although it is theoretically possible to cover such an area with energy-collecting devices and to convert the energy so collected into conventional electric power, the complexity of such a process, the vast quantity of metals and other materials required, and the astronomically high initial costs—when compared with the requirements for new steam or hydroelectric power plants—are great obstacles to such a vast new undertaking.

Recently, proposals have been made that space *satellites* be used as devices for converting solar energy to electricity (Fig. 3.23). These satellites would circle the earth at altitudes where they would receive almost constant sunlight. Solar cells would be developed to convert the radiant energy into electricity. The electricity would then be converted electronically in orbit to microwave energy with a wavelength suitable for penetrating clouds and would be beamed to earth where antenna-topped installations would convert it to electric power. One solar space satellite

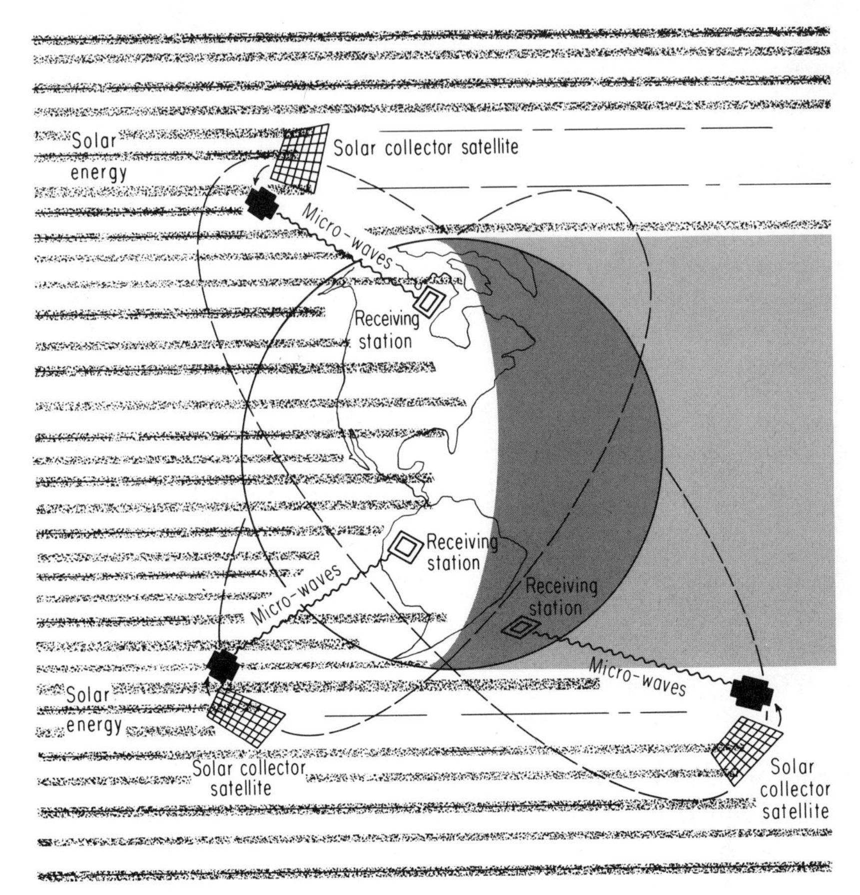

Figure 3.23 The sun is one of the most hopeful future sources of pollution-free energy for mankind. Solar-collector satellites circling the earth may someday transmit enough energy to terrestial receiving stations to provide for our growing power demands.

adequate to provide the 10,000 megawatts now demanded by the largest urban centers would require a rocket-launching system similar to those used for recent space vessel launchings.

Prospects: The costs of satellite-derived solar power are uncertain. A surface solar power plant producing 1000 megawatts would cost at least one billion dollars which is nearly four times the present cost of a nuclear power plant of the same output, but the ultimate cost of the electricity would probably be only about half a cent per kilowatt hour when the capital outlay is averaged over the life of the installation and when the very low operating costs are considered.

The direct use of solar energy would not produce the atmospheric pollution associated with combustion, and solar collectors would put a much smaller waste-heat load on the environment than fossil fuel or nuclear plants.

Much of the scientific knowledge required for large-scale use of direct solar energy already exists, but a very high capital investment must be anticipated in order to develop such power on a large scale. Unfortunately, these costs and the complicated technology make solar power unappealing to most energy-planners of today. Nevertheless, *the direct conversion of solar energy is the only really significant long-range alternative to nuclear power.*

THE LION'S SHARE

Energy production and use in the United States has for years been essentially unquestioned. In truth, growing use of electricity and other forms of energy is now regarded as a basic indicator of economic vitality and prosperity. But the times are changing: our landscape is crosscrossed with wires, and the pollution by-products of the power plants are beginning to affect the quality of our lives. There is no doubt that modern living requires some amount of external energy. The questions are: how much energy does the United States *need* to use and how much is it *reasonable* for us to use considering the limitations of the raw materials, the pollution by-products, and the need of still underdeveloped countries for energy to improve their living standards?

The United States now uses about 35 percent of the world's energy, yet it has less than six percent of the world's population. This disproportion will become larger because of the much greater rate of increase per-capita consumption of energy in this country than in the rest of the world and the rapid population growth of the underdeveloped countries. It is not impossible that the United States will soon be using half of the world's energy production!

Energy production in the vast underdeveloped regions of the world is rising very slowly. India, for example, has a nearly static, very low per-capita energy consumption. Its population growth largely cancels out benefits from any increase in total power production. If India and all the other underdeveloped nations in the world were able to reach by the year 2000 the standard of living of Americans today, the worldwide level of energy consumption would be 100 times the present world figures.

Approaching the Limits

We now face the need to supply energy for an ever-growing and power-hungry population from finite sources of that energy. Supplies of petroleum that once seemed vast may last only a few decades and all the world's coal no more than a century. Uranium, too, will likely last for but a short part of man's immediate future. There are also a very limited number of sites where dams can be built for water power, and those reservoirs built or to be built are destined to be destroyed by silt (Fig. 3.7).

It is not certain what fraction of all fossil fuels now has been used up. But even if we conservatively assume that we have already extracted only one percent of the earth's total reserve of all fossil fuels, the rate of extraction has been doubling about every 10 years. If this rate continues, we shall have extracted all of the remaining 99 percent of these fuels (coal will be the last) in less than 70 years from now.

Were the entire world to begin to consume energy at the same rate as that of the present-day United States, which would mean a consumption rate far more than 15 times greater than present world use, the resultant pollution produced by the number of additional power plants required would be so heavy that it is doubtful that men could long survive in such a damaged environment.

The Great Waste

The American population is rising at about one percent each year, but our use of energy increases by about five percent each year (Fig. 3.24). A large part of this annual rise is accounted for by our increasingly wasteful consumption habits.

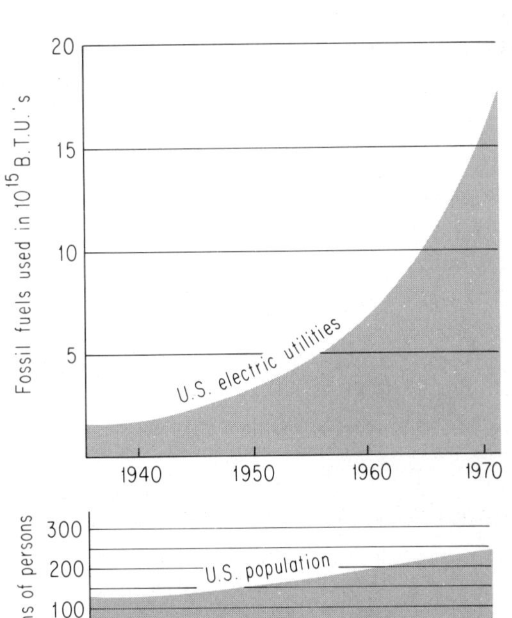

Figure 3.24 The generation of electricity by burning of fossil fuels in the United States is rising at a rate far in excess of population growth.

It is ironic that electricity, the most inefficient and environmentally costly form of energy, should also be the form most easily and commonly wasted. For example, we could never begin to estimate the amount of energy lost by leaving unneeded electric lights burning. How many Americans even concern themselves with turning off an electric light?

The dollar affluence of the United States has led to what may be the most resource-wasteful society the world has ever known. The kilowatt hours of electricity wasted in this country would be enough to electrify more than one underdeveloped nation.

Urban Energy Crisis: The cities of the United States are facing problems of energy production more directly than the less-populated regions. Much of the air pollution in cities is caused by the automobiles that bring commuters in and out and by the taxis and buses that cruise city streets. The major role of the internal combustion engine in causing air pollution was considered in Chapter 1. It seems clear now that private automobiles and perhaps all gasoline- and diesel-powered vehicles must soon be eliminated from central parts of the cities.

Many urban planners have advocated the exclusive use of electrically-powered cars on city streets. Unfortunately, the improvement in urban air quality that might be brought about by eliminating the internal combustion engine would have to be paid for by increased pollution from the power plants that supplied the electricity for the nightly recharging of the vehicles (Fig. 3.25).

The energy required to run electric cars and buses would be added to the increasing amounts of electricity being demanded by urban centers

Figure 3.25 Vehicles run by electrical motors require the generation of electricity, usually by burning coal or other pollution-producing fuels. Such vehicles are themselves free of air pollutants, but their efficiency is so low as to represent the production of large amounts of pollutants.

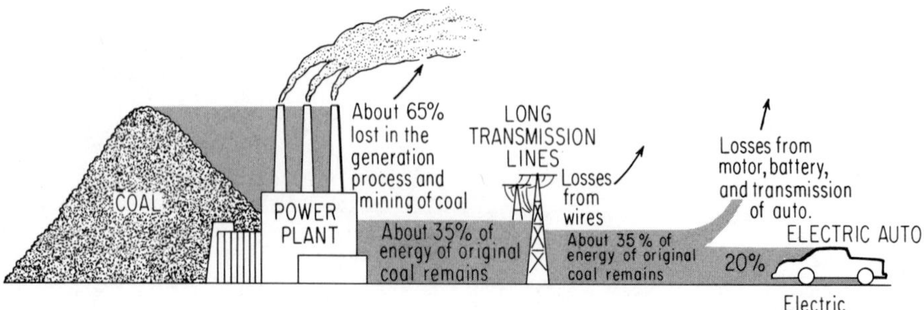

Figure 3.26 Electric transmission wires and support towers have become ever more numerous across the American landscape as electrical-power usage soars.

for lighting, air conditioning, heating, and other purposes. Power plants would have to be moved well outside of city limits, or there would be no overall improvement of urban air quality.

Conflicts often arise when new plants for urban power are located in outlying suburban areas. Another problem lies with increasing the distance between urban consumers and power plants; this is the necessity for long, high-voltage transmission lines that damage the landscape and necessitate the generation of all the more electricity to make up for the power losses of transmission through wires (Fig. 3.26).

A current example of the complex issues involved in widely separating the source of power from the user is the development of power plants in the Four-Corners region where New Mexico, Arizona, Utah, and Colorado meet. In this scenic area, much of it on Indian reservations of relatively low population density, large supplies of coal are being strip-mined to fuel the first two of six projected power plants (Fig. 3.27). These plants are intended to provide electricity for Los Angeles, Las Vegas, Phoneix, Albuquerque, and other cities all far distant from the Four-Corners.

The chief advantage to the consumers in the cities involved is more electricity without more pollution from urban power plants. But considerable protest has developed over the environmental effects of the Four-Corners plants. The first plant to go into operation, at Fruitland, New Mexico, was soon identified as the world's greatest single source of air pollution. The smoke plumes from the entire complex will undoubtedly affect the atmosphere over thousands of square miles of the United States (Fig. 3.28). Locally, the incomparable Four-Corners scenery, including Monument Valley, is becoming dimmer as the air pollution grows. Strip mines are progressively eating into the landscape. Water being extracted from below ground to use for steam and to allow transportation of coal in slurry form is likely to mean permanent changes for the mesa-dwelling Hopi Indians, long-dependent on ground water levels.

(a)

(b)

Figure 3.27 (a) Remains of a landscape after strip mining part of a Navajo Indian Reservation to supply electrical power to a number of cities like (b) Las Vegas, Nevada. (Photo courtesy of Las Vegas News Bureau.)

Energy Psychology

We are often told that it is necessary to increase the amount of energy we produce in order to "meet the demand." But such a statement ignores the environmental consequences of building more and larger power-generating plants.

It is assumed by power companies that the demand for power is real because people continue to use it in quantity. But the utilities and appliance manufacturers aggressively advertise new or "better" power-consuming devices, while there is no comparable advertising to point out the harmful effects of the increasing energy consumption.

(a)

(b)

Figure 3.28 (a) The first of a series of giant plants being built in the Four-Corners region of the American Southwest. The pollutants from this plant near Farmington, New Mexico spread over hundreds of miles of landscape. (Figure 3.27a is a photo of coal source for this plant.) (U.S. Department of Interior, Bureau of Land Reclamation.) (b) Pollution from this power plant blankets the surrounding Navajo Indian Reservation. Note Navajo homes in foreground.

Because electricity is the least efficient and most environment-damaging form of energy, we should seriously consider the methods available to reduce the present rate of consumption. One of the most obvious approaches is to require power companies to reverse the present rate structure that rewards heavy consumers of electricity with lower rates per kilowatt hour. In Norway, for example, a low charge is made to domestic consumers until their consumption rises above an established, reasonable level after which the cost per kilowatt hour *increases.*

It may be unrealistic to expect the power companies and appliance manufacturers in the United States to set limits or discourage overuse of electrical and other forms of energy. The limits may have to be established by enlightened individual citizens, and their consumer groups, and legislative representatives. But limits *must* soon be set.

Each day we are more clearly confronted with the penalties for our nation's long-uncontrolled appetite for the products of energy and resources. Having avoided the signs and warnings for decades, we are finally awakening to a gray world of polluted air and water, crisscrossed by highways, strung with power lines, and scarred by strip mines. And there are several billion more people, in the underdeveloped nations, ready to follow the same course in the coming years.

Suggested Readings

AVERITT, P., *Coal Resources of the United States,* Jan. 1, 1967, *U.S. Geological Survey Bul.* 1275, 1969.

COOK, EARL, "The Flow of Energy in an Industrial Society," *Scientific American,* Vol. 224, No. 3, 1971.

CURTIS, RICHARD and ELIZABETH HOGAN, *Perils of the Peaceful Atom: The Myth of Safe Nuclear Power Plants,* Doubleday, New York, 1969.

DANIELS, F., *Direct Use of the Sun's Energy,* Yale University Press, New Haven, Conn., 1964.

GOFMAN, JOHN and ARTHUR RAMPLIN, "Radiation: the invisible casualties," *Environment,* Vol. 2, No. 3, 1970.

HUBBERT, M. KING, "The Energy Resources of the Earth," *Scientific American,* Vol. 224, No. 3, 1971.

ZAPP, A. D., "Future petroleum producing capacity in the United States," *U.S. Geological Survey Bul.* 1142-H, 1962.

4
Oceans

Union Pacific Railroad Photo.

The world ocean covers about 70 percent of the earth's surface. The great expanses of the Pacific, Atlantic, Arctic, and Indian Oceans, along with the various seas, gulfs, straits, and bays are all interconnected and share a very similar composition.

Most of the earth's elements are present in sea water, at least in minute quantities, including gold and other precious elements. The famous futile attempt of a post-World War I German chemist to extract gold from sea water in order to pay off the German reparation payments was a curious proof that, although, almost all the elements are present in the sea, most are in concentrations much too minute to be extracted on a commercial basis.

Sea water is a solution of various salts, sodium chloride being the most abundant (about 86 percent) of the salts. Others present, in order of decreasing abundance include: magnesium chloride, sodium sulfate, calcium chloride, and potassium sulfate. In addition to salts, seawater contains small amounts of dissolved gases, such as oxygen and carbon dioxide. Much of this content is derived from the atmosphere during stirring of the surface by winds, and some is produced by the life present in the oceans.

The natural composition of the sea has been gradually altered by the addition of man-made wastes, chemicals, and pollutants of various kinds. The degree of man's alteration of oceanic chemistry and its environmental significance will be considered later in this chapter.

DEPTHS OF THE SEA

It was in 1872 that the British ship *Challenger* set out on a great, systematic expedition to explore the deep seas of the world and was able to find some of the deepest regions of the western Pacific. For more than half a century after the Challenger Expedition, few depth determinations were added to our knowledge of the deeper oceans. With so few actually measured depths, scientists assumed that the ocean bottom was a nearly level plain.

Methods of oceanic exploration changed drastically after World War II when the war-developed echo sounder became available to science. With this device, sound impulses sent from a ship are sent back as echoes off the sea floor, and the time elapsed immediately indicates the depth. A continuous sounding by a moving ship produces a graphic recording of the profile of the sea floor and all its features.

The Sea Floor

The floor of the ocean can be divided into three major regions. First, the continental shelf, a shallow, life-rich marginal zone between land and deeper sea. Adjacent to the shelf is the continental slope where the continents actually end and the sea bottom slopes down two miles or more to the true floor of the deep-ocean basin (Fig. 4.1).

The continental shelves border most of the earth's coastlines, miles wide and sloping very gently away from the shore (Fig. 4.2) to a depth of about 500 feet. They were built up largely from sediments carried out to sea by rivers, deposited along the shoulders of the continental blocks, and planed smooth by the action of waves and currents.

Off some coasts, especially in the Atlantic, the shelf is quite wide, reaching 100 miles or more before dropping down toward the deeper

Figure 4.1 Simplified diagram of an ocean showing the shallow shelf, the slope, and the deeper basin.

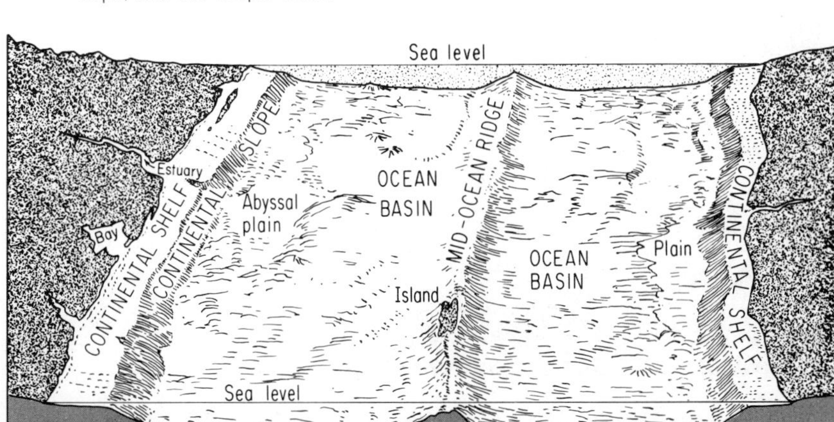

(a)

(b)

Figure 4.2 (a) The innermost continental shelf exposed at lowest tide. This is a region highly susceptible to pollution carried out from the land. Japanese coast. (Photo courtesy of Japan Tourist Association.) (b) High tide covers the entire continental shelf. (Photo by author.)

ocean. Off young, mountainous shores, like the Pacific coast of the Americas, there has been little time for erosion and sedimentation, and the continental shelves were very narrow.

At the outer edge of the shelves, the continental slopes drop off into the depths at inclinations as great as 300 feet per mile. The deep-ocean or basin floor begins at the foot of the continental slopes. Most of the sea floor, about $\frac{5}{7}$ of the total expanse of the ocean, is within the basin region which averages about $2\frac{1}{2}$ miles in depth. It is in this vast stretch of darkness that oceanographers have been making the most significant and surprising discoveries about the nature of the sea floor.

Probably the greatest single geographical discovery made in modern times was that of the mid-oceanic ridge, a great mountain range 40,000 miles long, snaking its way through every ocean of the world. Its presence was first indicated by the Challenger discovery of a rise in the middle of the Atlantic. But the significance and great length of this mountainous rise went unrecognized until echo sounders came into use after World War II. We now know that the great undersea ridge runs right around the world, has a number of branches, and that a fracture zone is to be found at the center of the entire ridge. Most of the earthquakes that take place under the sea are centered along this zone. The oceanic ridge and fracture zone can be traced onto the shore in places as in eastern Africa where it becomes the Great Rift Valley.

It was once assumed that the deepest levels of the seas would be found farthest from land. But we now know that the deepest parts of the ocean are all near land, mostly around the margins of the Pacific. These are the deep ocean trenches, most of them over 30,000 feet deep. The deepest of these is the Mariana Trench (south of Japan, near Guam), over 36,000 miles deep, into which two men descended in a bathyscaphe in 1960.

Relatively few parts of the deep sea floor have been found to be flat. Those that do exist are called abyssal plains, formed mainly by blanket-deposits of sediment carried out by turbidity currents. Volcanoes have been found by the hundreds in the deep sea, rising as towering cones from many parts of the ocean floor, occasionally reaching high enough to jut up above sea level, as in the Hawaiian Islands.

CIRCULATING WATERS

Surface Movements

Like the air, most of the sea is in motion. Propelled by temperature differences, winds, and other factors, the waters of the ocean spread out from one region to another, forming the currents or streams long familiar to mariners and geographers. On a planetary scale, there is a tendency for the cold, denser waters of the high latitudes to sink and spread gradually equatorward under the warmer, lighter waters of the lower latitudes and for the *surface* waters at the equatorial, low latitudes to spread poleward. But this general tendency is greatly modified by the dragging effect of prevailing winds, the influence of irregular shorelines, and the rotation of the earth.

Prevailing winds, such as the middle latitude westerlies, exert a strong, unidirectional drag on vast stretches of sea water and produce the major surface currents, the great *drift currents* of the oceans. Together with currents set up by density differences, these drift currents develop the continuity of ocean circulation.

Ocean currents caused by wind drag do not flow in exactly the same direction as the wind itself. The rotation of the planet under the moving sea results in a veering to the right of currents in the Northern Hemisphere and to the left in the Southern Hemisphere (the Coriolis effect). Thus, surface currents flow in directions between 20 and 45 degrees to the right of the prevailing wind directions (Northern Hemisphere); a prevailing westerly wind (one blowing due east) may cause a surface current flowing toward east-southeast.

World Currents

In the Atlantic Ocean, there are two great circular patterns of surface water motion called *gyres,* one centered about 30 degrees north latitude and one centered about 30 degrees south latitude. Each gyre is generated by two sets of prevailing winds, the tropical easterlies, which bound both gyres on the equatorial side and the prevailing westerlies, which blow along the north side of the northern gyre and along the south side of the southern gyre, and by the Coriolis effect on the moving waters (Fig. 4.3). A similar but somewhat more complex surface circulation exists in the Pacific Ocean. A simple, counterclockwise surface circulation exists in the mostly land-surrounded Arctic Ocean.

In the Atlantic Ocean, water swept westward by the easterlies moves into the Caribbean region, turning right, ever more northward, then

Figure 4.3 Opposite page. (a) Diagram of circulation patterns in an idealized ocean of simple shape. (1) Surface currents caused by prevailing winds (see Chapter 1), differential heating, and Coriolis effect. (2) Cross section shows rising and spreading of waters away from warm, equatorial regions, sinking and returning flow from colder, polar regions. (b) Actual surface currents in and near the Atlantic Ocean reflects the irregularity of the surrounding continents as well as the factors illustrated by (a) above.

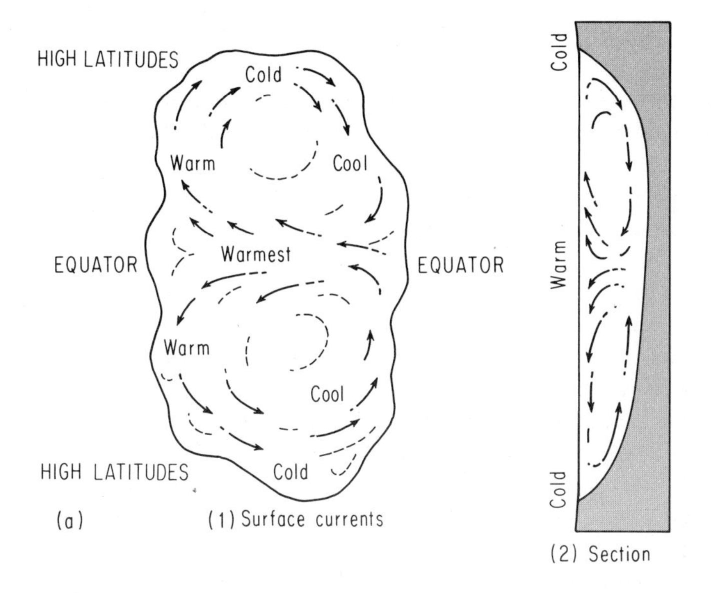
HIGH LATITUDES
Cold
Warm
Cool
EQUATOR
Warmest
EQUATOR
Warm
Cool
HIGH LATITUDES
Cold
(a)
(1) Surface currents
Cold
Warm
Cold
(2) Section

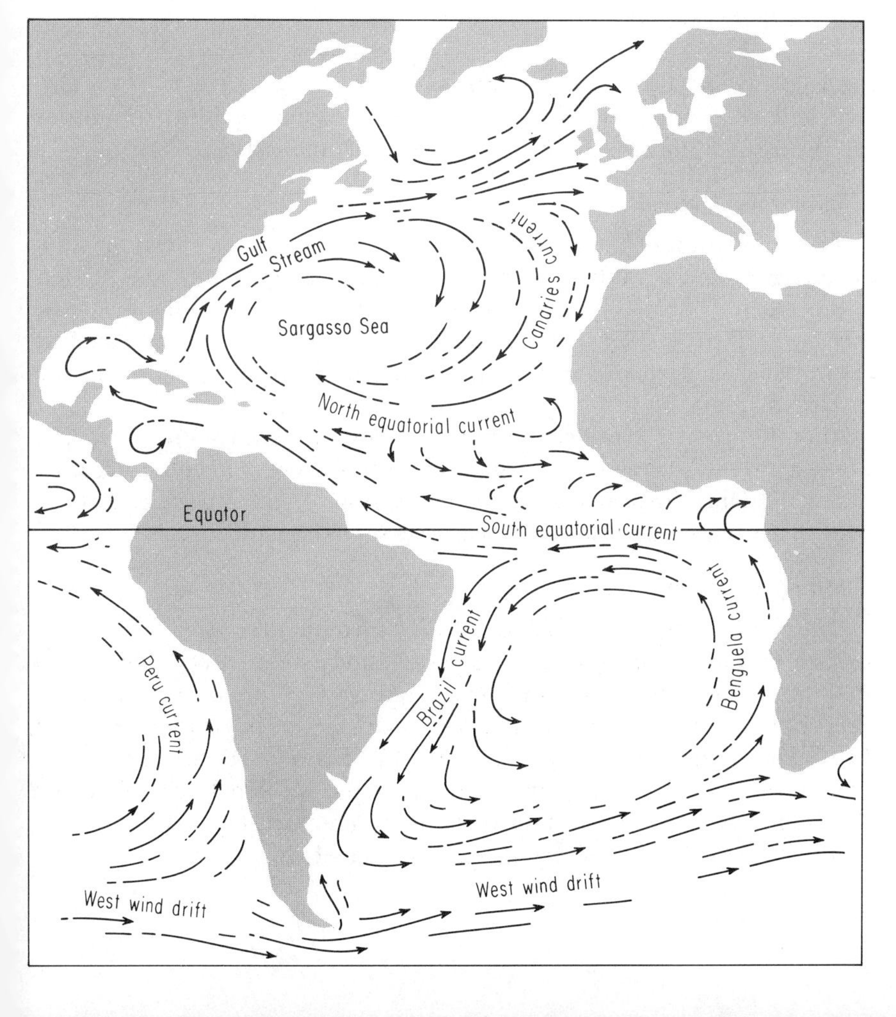
Gulf
Stream
Sargasso Sea
Canaries current
North equatorial current
Equator
South equatorial current
Benguela current
Brazil current
Peru current
West wind drift
West wind drift

toward the northeast along the coast of North America. This is the well-known Gulf Stream, first plotted by Benjamin Franklin, and reaching speeds of three miles an hour in places. The westerly winds of the North Atlantic drag these water toward the European continent where the flow divides, the North Atlantic current flowing north-northeast along Scandinavia and into the Antarctic. Some of the Gulf Stream flow, turned southward by the Coriolis effect, becomes the Canaries current, finally reaching equatorial latitudes where the drag of the easterlies helps produce current flowing back (westward) toward the Caribbean, thus completing the great gyre of the North Atlantic at the center of which are the relatively stationary waters of the Sargasso Sea (Fig. 4.3b).

The flow of surface waters in the South Pacific describes a gyre of especially vast proportions. Surface waters here are propelled by the equatorial easterlies on the north and by the strong westerlies that blow all around the planet at about 50 degrees south latitude, and are ever-directed by the Coriolis effect in that hemisphere. A segment of this vast South Pacific gyre is the Peru or Humboldt current, the water of which moves north along the western coast of South America and then veers northwestward to become the south equatorial currents. This current will be considered below in its role in supplying ideal conditions for the great numbers of fish being caught off the coast of Peru.

Subsurface Currents

The circulation of the deeper ocean and the directions and intensities of currents flowing well beneath the surface are not, yet, well understood and only sketchily mapped. The Cromwell current, for example, was first traced in 1952. This is a fast equatorial flow concentrated at about 300 feet deep in the Pacific, flowing opposite in direction to the surface currents.

Another current has been discovered flowing southwestward in the western Atlantic between 8000 and 10,000 feet deep. This flow has been called the Gulf Stream countercurrent because of its apparent relationship to the oppositely-directed surface flow of the Gulf Stream.

Mixing of the Waters

One result of the numerous drifts and currents that spread across the surface of the ocean and the still-poorly understood deeper flows is a slow but steady mixing of the world's sea water. Bottom waters in the deep oceans return to the surface in about one or two thousand years. But because the oceans have existed for billions of years, we can assume that their waters have been mixed more than a million times. Thus, we should not be surprised to find that uncontaminated sea water is nearly identical, regardless of where samples are collected.

OCEAN POLLUTION

Until very recently the ocean has been thought of, by laymen and experts alike, as impervious to all intrusions, vast enough to dilute, absorb, and eliminate all wastes and debris that could ever be produced by man. But in the 1960s it became evident that this concept of infinite capacity was not quite true. Although it covers 140 million square miles of our planet, the ocean has definite limits and is no longer completely effective in coping with the ever-increasing quantity of sewage and waste or with the novel new chemicals that pour off the land from our cities and factories. At the moment, it is along the coast lines, where the wastes of our civilization are discharged, that the natural chemical composition of sea water is being most noticeably altered and the life of the ocean most seriously imperiled.

Fragile Coasts

The ocean is much like a giant bowl with a wide, slightly depressed brim and full of water. The barely-flooded brim consists of the continental shelves. These shallow parts of the sea penetrate into the margins of the land as bays, sounds, and estuaries, comprising about $\frac{1}{10}$ of total area covered by the seas. Here develop the sand bars, barrier beaches, and reefs, features which bar the ready flow of river waters into the deeper sea and of ocean currents onto the shore, thereby increasing the concentration of pollutants along the coast.

Onto the shallow continental shelves is dumped almost all of the garbage, sludge, and other wastes disposed of in the oceans. Here concentrates much of the sewage and chemical wastes washed out by river waters from households, chemical plants, factories, oil refineries, and hospitals.

As seen from a ship, the most obvious coastal pollution takes the form of floating plastic containers and objects, such as bleach jugs, cups and toys, styrofoam articles, and bottles of all kinds (Fig. 4.4). This type of

Figure 4.4 Pollution of coastal waters often takes the form of floating debris. Coast near New York City. (Photo by author.)

debris may float for years. The water is commonly stained red, yellow, or brown by the discharges of manufacturing plants. A closer look at the surface often reveals human sewage and bits of tissue paper.

On the sea floor along coastal regions, it is common to find discarded debris in such quantities that dragging for bottom fish can be a frustrating or even dangerous experience. Nets are frequently torn by junk cars and other metal objects or snagged on the live shells, bombs, or poisonous gas canisters dumped by the military during decades of operations.

Rivers are the major sewer pipes contributing to the pollution of the coastlines. Any river that flows through an urban, industrial area is full of pollutants. Thus, in many ways, coastal pollution is a by-product of river pollution (Fig. 4.5). But coastal and harborside industries also add enormous quantities of polluting substances directly to the seas. Most of our largest cities are located along the coast and generally pump or carry their partly treated or untreated sewage directly into harbors or to outfalls a short distance offshore. Some communities carry sewage sludge by barge to dumping grounds several miles from the coast, but still in the relatively shallow waters of the continental shelf.

Many estuaries, harbors, and bays have become little better than sumps for waste. In these regions, the effects of growing coastal pollution are obvious. Swimming is forbidden at many beaches. Shellfishing declines

Figure 4.5 Industrial waste in the Hudson River on its way to the Atlantic Ocean. (USDA Forest Service.)

or must be ended as infectious hepatitis germs are concentrated by these filter-feeders at levels that make the bivalves a hazard for human consumption. Shellfish in coastal waters have been found with polio virus concentrated at a level over 50 times greater than that of the surrounding water. By 1971, a quarter of the shellfish beds along Canada's Atlantic coast had been closed because of dangerous contamination. In many coastal regions, lobsters, crabs, and fish that do manage to survive for long in water heavily polluted by sewage suffer from cancer-like growths and other diseases.

The growing number of nuclear power plants being constructed along coastlines may mean a double problem of radioactive isotope contaminations and thermal pollution of shallow waters (see Chap. 3), the hazards of which are just now beginning to be perceived.

Gulf Coast: Most of the Gulf of Mexico is relatively shallow and well on its way to becoming dangerously polluted. Discharges of urban sewage and industrial wastes enter the gulf from a number of sources: The Mississippi makes a major contribution with its gigantic discharge of water, sediment, and pollutants. Various cities in the United States, and Mexico, and islands like Jamaica and Cuba add their wastes. Cellulose fibers, which can only be the remains of toilet paper, abound in the waters between Florida and Cuba. Cellulose is a substance that is essentially nonbiodegradable and an extremely serious component of marine pollution in the quantities being produced today.

The Gulf of Mexico may soon become a giant cesspool of civilization. Semi-enclosed by lands or islands, the gulf has too many bays and harbors emptying into it. Galveston Bay, Texas, for example, slowly flushes its waters into the Gulf of Mexico. The Houston Ship Channel, which connects the bay with Houston, is lined with industries pouring waste equivalent to the raw sewage of over two million people into Galveston Bay. Oil refineries, petrochemical plants, and paper mills all add their processing wastes to the bay; the adjacent city of Galveston has been discharging $1\frac{1}{2}$ million gallons of its own sewage into the gulf each day.

Galveston Bay was once a major shellfish-producing area but is now closed to oyster harvesting because of pollution. It has also been a vital nursery for shrimp and various fish so that the growing contamination of these and other breeding grounds threatens to damage the entire fishing industry of the Gulf of Mexico.

Atlantic Coasts: In some regions, harbors and bays have become so polluted that cities have been carrying sewage and waste out of these water bodies and into the ocean before dumping (Fig. 4.6). This practice has already created a number of major offshore centers or sinks of

Figure 4.6 Many cities dump their garbage at sea. These barges of waste from a city along the Hudson River are bound for a dumping ground (pollution sink) off the mouth of the Hudson. (USDA Forest Service.)

pollution. A notable example is New York City's "dead sea," a dumping ground largely devoid of marine life, located some nine miles outside of New York Harbor. Another is located off Cape May, New Jersey. These pollution sinks have been growing in size at a rapid rate as new regulations against river pollution are put into effect, (or old laws enforced), so that more and more rivershore cities ship their wastes out to sea by barge.

In 1972, New Jersey's shore communities were restrained from their practice of regularly emptying sewage sludge holding tanks directly into the sea just 1000 feet offshore. These wastes may now have to be carried out by barge and are likely to add to the proportions of the "dead seas" off New York and Cape May.

Europe: Across the Atlantic, the Baltic Sea is a shallow, estuarine branch of the world ocean. The Baltic is surrounded by industrialized nations, including Sweden and Russia. The buildup of mercury in Swedish waters is now high enough to exceed limits proposed by the World Health Organization. DDT and phosphate input are both high enough in the Baltic to be of concern.

Deeper basins in the Baltic have accumulated waste to the point of deoxygenation, and production of poisonous hydrogen sulfide has been detected. Swedish pulp mills have been pouring in waste cellulose fiber that fails to decompose and does great damage to bottom life. Largely enclosed by land, cut off from the currents of the Atlantic, and capable only of very slow flushing into that ocean, the Baltic may be even closer to becoming a giant industrial-urban cesspool than the Gulf of Mexico. (Fig. 4.7).

The Mediterranean is also a land-locked sea. Portions of its coastline, especially along France and Italy, are becoming heavily polluted. The problem has been most obvious at places where there is heavy industrialization or urbanization, like Marseilles, Nice, and Fos. In the waters off Fos, France, nine out of 13 species of fish once used for food have disappeared, and the remaining species are scarce.

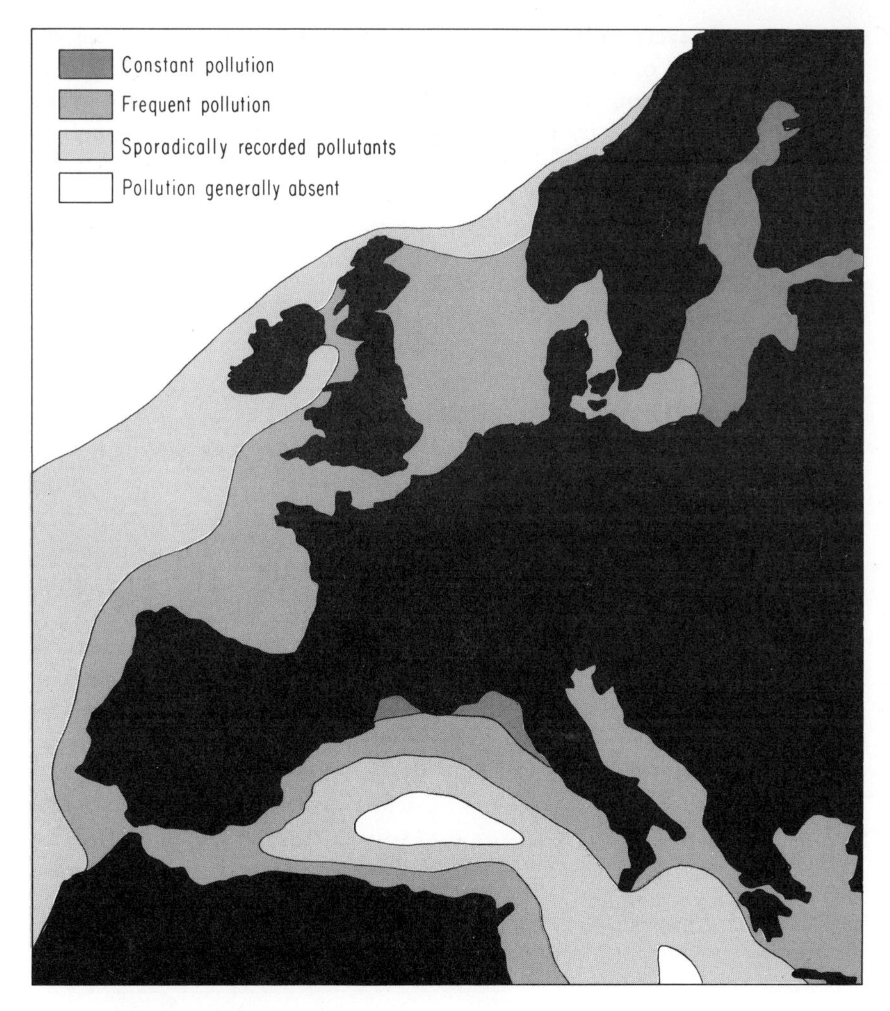

Figure 4.7 Coastal waters of Europe are becoming ever more polluted. This diagram shows the approximate pattern of pollution levels around Europe in 1972.

On a scale still larger than the Mediterranean, Baltic, or Gulf of Mexico, we may consider the North Atlantic itself. As vast as it is, the North Atlantic receives a very disproportionate share of the world's pollution. Into its waters empty dozens of major rivers draining the industrialized nations that border it on both sides. Among the most polluted of these are the Rhine, Seine, Rhone, Tiber, Hudson, and Delaware.

The continuity of the North Atlantic with the South Atlantic and its connection to other parts of the world ocean might seem to minimize the danger of severe pollution, but there are strong indications that the rate of discharge of waste substances has been far exceeding this sea's ability to break them down. Thor Heyerdahl's recent historic crossing of the North Atlantic by primitive papyrus vessel was slow enough to reveal the continuous spread of floating pollutants from one side of the ocean to the other. Buildup of oil waste is especially obvious in the middle region of the North Atlantic where the patterns of surface currents have produced the relatively still waters of the Sargasso Sea. This phenomenon is considered later in this chapter.

Pesticides and Related Chemicals

DDT: The well-known pesticide DDT was developed less than 30 years ago. After its early uses in combating body lice and typhus during World War II, it became spectacularly successful as a weapon against insects: in public health by controlling malaria, which is spread by mosquitoes, and in agriculture by decreasing losses from the various insects that damage crops. But in the past quarter century, this insect-killing poison has also thoroughly permeated our environment.

DDT is now absorbed in the fatty tissues of almost all animals, including the author and the reader. The average American had over 10 parts per million of DDT in his body fat in 1971. Most Europeans had less, but Israelis averaged nearly 20 parts per million; even Eskimos living in remote northern regions had over two or three parts per million.

DDT is spread over the planet by wind and, to a lesser extent, by flowing waters (Fig. 4.8). When sprayed into the air over farmland, about

Figure 4.8 Pesticides, such as DDT, are carried into the oceans via rivers and wind. The ultimate source is spraying for control of insects and other pests.

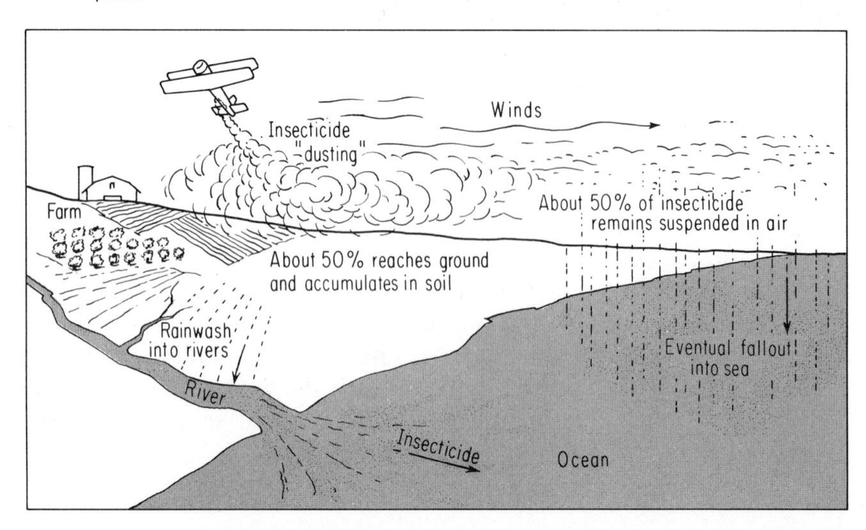

half of the pesticide becomes suspended as small particles or crystals that tend to be picked up by even the gentlest winds, much as pollen grains are widely dispersed, and brought down gradually into the seas by rainfall.

DDT is only slightly soluble in water, but it is an extremely stable compound that requires as much as 50 years to lose half its strength. As organisms in the sea absorb the substance, they make room for more DDT to be dissolved in the water. Plant and animal life in the sea constitute a chain that concentrates DDT at spectacular levels (Fig. 4.9).

Figure 4.9 (a) Oceanic food chain causes a cumulative build up of DDT, mercury, and other pollutant substances in marine life forms. (b) As pollutant particles enter the sea, they dissolve and can be assimilated by organisms, or they come to rest on the bottom from which they can be later dissolved and added to the food chain.

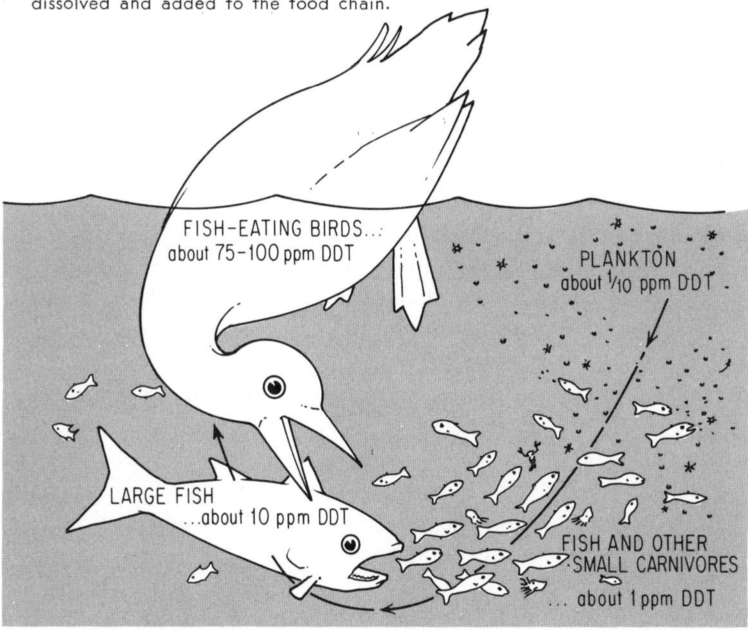

(a)

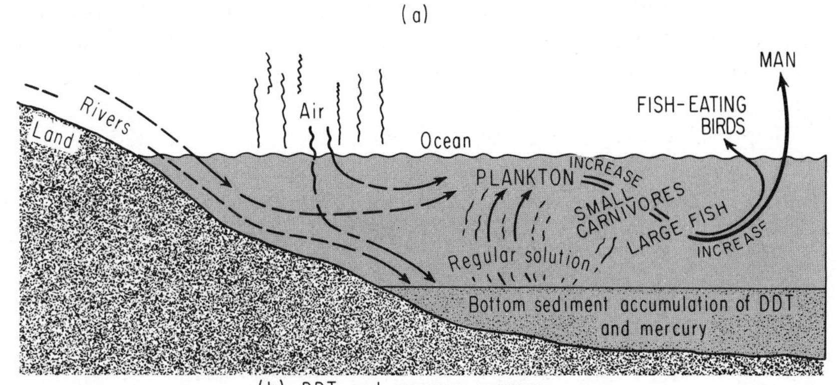

(b) DDT and mercury sources

Plant plankton, at one end of the chain, may contain only a fraction of one percent of one part per million, but the small fish that feed on the plants commonly build up a concentration of over one part per million. Carnivorous fish that are farther along the food chain may contain over 10 parts per million, and fish-eating birds may reach levels of 75 to 100 parts per million of DDT. Thus a community may concentrate the pesticide by a factor of more than 1000 times the level dissolved in the sea around them.

As DDT accumulates in greater and greater doses, it alters various biologic processes and causes injurious changes to plants and animals along the food chain. These changes range from inhibiting growth and reproduction of the phytoplankton, basic foodstuffs of the sea, to causing the failure of reproduction in fish-eating birds. In the eggs of some of these birds, DDT residues as high as 2500 parts per million have been found. DDT interferes with the mobilization of calcium in the oviduct so that eggs may be laid with little or no shell. There are signs of egg-shell thinning throughout the world. In Long Island Sound, New York, many of the terns had lost their ability to reproduce by 1971, and most of the young that did manage to hatch had gross deformities such as crossed beaks and unequal limbs.

Humans who habitually eat substantial amounts of fish may develop an above average concentration of DDT with the passage of time. Studies of the effects of the cumulative buildup of DDT and other pesticides on man are quite new, and results are being closely watched by environmentalists.

DDT was outlawed for most purposes in the United States in 1972, but its use is still rising in many of the underdeveloped countries. Even if the use of DDT in the world were to cease immediately, the concentration in the sea would continue to build up as suspended particles slowly settle down from the air, and the long-lived residues in soils are gradually carried out by rain and rivers.

Other Chlorinated Hydrocarbons: Chlorated hydrocarbons other than DDT are also becoming widespread in the environment; dieldrin is an especially deadly example. They are used as insecticides, fungicides, and herbicides. The biological side effects of most have not been adequately studied.

The polychlorinated biphenyls (PCBs) are chemicals related to DDT which are used extensively in industry as plasticizers. They can be quite toxic and are now extremely widespread in our environment. PCBs are used in vast quantities of disposable plastic goods. Combustion temperatures in commercial incinerators volatilize rather than destroy these chemicals so that they are slowly released to the air and find their way into the ocean in a variety of ways and are concentrated in the same manner as insecticides. The deformed terns of Long Island Sound were found to contain PCB along with DDT.

The chlorinated hydrocarbons are unique, man-made compounds differing greatly from organic material that is more readily decomposed by bacterial action. Neither bacteria nor other organisms have evolved that are capable of rapidly breaking down the chlorinated hydrocarbons. As a consequence, they persist in the environment for long periods of time. It is the combination of persistence, foodchain concentration, and toxicity that has caused the growing fear of the continued use of these substances.

A number of insecticides are available as substitutes for the chlorinated hydrocarbons that are less persistent in the environment, being quickly decomposed, and less harmful. But their lack of persistence means that more frequent application is necessary to achieve the same degree of insect destruction. Use of these substitutes tends to be more expensive than DDT, and some may not be completely safe.

The use of chlorinated hydrocarbon pesticides is likely to be ended soon in most of the developed countries. But the need for more food in underdeveloped countries with rapidly rising populations will probably prolong present use in these nations until adequate substitute insect pest control methods are developed and made widely available.

Toxic Metals

Another type of ocean pollution consists of metals such as copper, mercury, and lead which have been found concentrated in large amounts by marine organisms which are not able to eliminate the toxic elements from their systems except over long periods of time.

Copper: Copper concentrations of about three parts per billion occur naturally in the sea and are harmless, but $\frac{1}{10}$ of a part per million is toxic to most marine life, and one part per million can kill clams and other marine life and greatly slow down the ability of the kelp variety of algae to produce oxygen and food.

The Japanese "green oyster" phenomenon results from high concentrations of copper borne to the sea by rivers polluted by copper refineries. Oysters have been killed in great numbers where the rivers empty into the sea, and green-colored oysters can be found for miles around.

Mercury: Mercury pollution of the oceans has been in the news since the detection in 1970 of high concentrations of the metal in tuna and swordfish.

Mercury's toxicity is the result of its ability to disrupt the central nervous system. Long known to be poisonous, the liquid metal, nevertheless, had been assumed to be harmless when discarded into rivers or seas because of its tendency to settle to the bottom along with sediment. But the true danger lies in *methyl* mercury, a substance now known to be produced from the bottom mercury through the action of microorganisms. Methyl mercury can be assimilated by aquatic life, being concentrated at higher levels by each organism along the food chain (Fig. 4.9). The large fish, like tuna, have concentrations potentially dangerous to man, who stands at the end of the chain eating the fish.

The most dramatic example of human mercury poisoning from eating contaminated marine life, in this case shellfish, occurred between 1953 and 1960 to the inhabitants of villages on Minamata Bay on the coast of Kyushu, Japan. Over 100 cases of severe poisoning or death featured the bizarre symptoms and mental deterioration of mercury poisoning. The mercury was finally traced to a plastic-manufacturing plant dumping mercuric chloride into the bay.

In recent years, recognition of industrial sources of mercury pollution has been rapidly increasing. Mercury is widely used including in electronic industries, chlorine manufacturing plants, the paper and pulp industries, and in plastic manufacturing. All of these can emit metallic mercury or mercuric compounds that are discharged as effluent waste into waterways where they can become altered to mercury. It can also be carried into the atmosphere as mercury vapor from any refineries or other activities that heat mercury or other ores or compounds containing the metal. Mercury compounds have also been used as pesticides and fungicides in agriculture.

Some Baltic coastal regions have become so contaminated with mercury that the sale of fish caught there has been prohibited. Parts of Lake Erie have been yielding mercury-contaminated fish, and the fish and waters of 20 states are known to be contaminated with the toxic metal. The widespread mercury content of fresh water fish has led to fears that the ocean will contine to rise in its mercury content.

Lead: Lead is another example of a toxic, heavy metal that has been increasing rapidly in our environment. Tetraethyl lead has been added to gasoline for several decades to improve the performance of the fuel in automobile engines and is finally discharged into the atmosphere as the gasoline is burned (see Chap. 1). Lead particles can be carried great distances by the wind, finally settling (or washed down by rain) into the seas.

The progressive contamination of the atmosphere by lead is well-documented by studies of concentrations in polar snow layers going back hundreds of years. A very gradual increase in lead content occurred between 1750 and 1924, followed by a very rapid increase after the introduction of leaded anti-knock gasoline in the 1920s.

The settling of lead particles out of the air is verified by observations of greater lead concentrations in the sea in surface waters than in deeper waters and also by clearly higher amounts of lead in waters adjacent to most heavily populated regions. This reflects the high use of gasoline on lands nearby. We still know very little about the toxicity of the lead that is building up in the open seas.

OIL ON THE SEAS

Oil is rapidly becoming one of the most widespread pollutants in the ocean and may well be the most dangerous marine contaminant of all (Fig. 4.10).

Danger of Shipping Oil by Sea

In 1967, the oil tanker Torrey Canyon ran aground on shallowly submerged rocks off the coast of Great Britain. Millions of gallons of crude oil from faraway Kuwait gushed out, spreading over a hundred square miles of ocean in a few days. The great slick lapped onto the shores and beaches of Cornwall and Brittany. Mile after mile of coast was coated with evil-smelling oil, grounding and killing birds by the thousands, untold numbers of oysters, and other marine life. Measures taken against the oil by desperate Britons included unsuccessful burning attempts and multiple assaults with detergents that were more catastrophic to life than the crude oil had been. Over 5700 birds were washed clean of oil by well-meaning rescuers but all except 100 of these were soon dead.

The Torrey Canyon carried 118,000 tons of crude oil. Many tankers have larger capacities (Fig. 4.11). The largest tanker afloat in 1971 had

Figure 4.10 Sources of marine oil pollution.

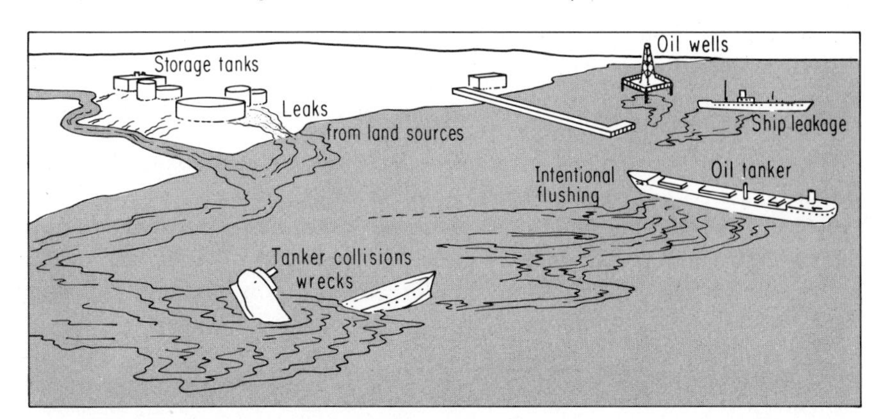

Figure 4.11 Giant tanker carrying petroleum products on the high seas. (Courtesy of Mobil Oil Corporation.)

a 300,000 ton capacity. Tankers of 500,000 tons were soon to be constructed, and 800,000 ton vessels were being planned, with oil tankers of one million tons on the drawing boards in more than one country.

The danger of collision of giant tankers can be imagined when we consider that it requires one-half hour for a 250,000 ton ship traveling at full speed to come to a halt! A wreck of one of these supertankers would make the *Torrey Canyon* spill look small indeed.

During the single year of 1969, 200 ships carrying oil cargoes were damaged or wrecked through collisions or other accidents. Thousands of oil spills occur each year through such oil tanker mishaps, by intentional or accidental ship discharges, and from leaks at sea. Oil spills from tanker accidents are spectacular, sometimes well-publicized events, but they are responsible for only about 10 percent of the total amount of oil entering the seas. The remaining 90 percent of the petroleum comes from *normal* operation of tankers and other vessels powered by oil, as well as from offshore wells (see below), refinery operations, and discharge of oil waste from the land.

Danger of Drilling Holes in the Sea Floor

The great blowout of a Union Oil well in the Pacific off Santa Barbara, California, in 1969, spread deadly crude oil over 30 miles of California coast. Entire plant and animal communities in the intertidal zone at Santa Barbara were killed by a layer of encrusting oil that was nearly an inch thick. As cleanup attempts proceeded, it soon became impossible to distinguish between the effects of the oil itself and those of the detergents used to try to control the pollution.

Leaking oil wells have become so common that a well that began spewing petroleum into the Gulf of Mexico in 1971 was hardly considered worth writing about by most of the news media.

At present, offshore production accounts for over 15 percent of total crude oil production, with about 9000 oil wells operating off the American coasts in 1971 (Fig. 4.12). The percentage and number of wells may be expected to increase greatly as new underwater fields are discovered (see Chap. 3).

Because of the hazardous conditions under which offshore drilling is generally conducted and the plans for expansion of drilling operations, geologists have predicted the probability of new spills off Santa Barbara. Attempts have been made to impose a moratorium on oil drilling in certain offshore areas of highest risk, like the Santa Barbara waters, and to bar new exploratory drilling operations in the Atlantic, until there is a greater guarantee of safety than that presently given by the oil companies.

In order to carry crude oil from offshore wells to shore refineries, there are tens of thousands of miles of pipeline, much of it rusting and in danger of breaking and capable of releasing quantities of crude oil in environmental disasters that could easily eclipse the spill at Santa Barbara.

Figure 4.12 An oil drilling rig located in the Gulf of Mexico. (Courtesy of Mobil Oil Corporation.)

Other Sources of Oil Pollution

Another way that oil finds its way to sea is the procedure in which oil tankers fill their tanks with seawater after discharging the oil cargoes, in order to maintain stability on their return voyages. Usually the oil-contaminated ballast water is drained into the ocean, and the tanks are cleaned before entering the port of loading. By international agreement some regulations have been developed to reduce this source of oil pollution. Generally, such discharges are supposed to take place at least 50 miles from shore, but methods of enforcement are inadequate, and the agreement is not always a practical safeguard.

Ordinary tanker operations added over 500,000 tons of oil to the sea in the single year of 1969. Nor are tankers the only offenders. Indeed, all seagoing ships pump bilges, which invariably contain some oil, into the ocean.

Normal refinery and well operations add thousands of gallons of oil to rivers and oceans each day (Fig. 4.13). This includes both discharged oil wastes and tank leakage. As much as a million tons of automotive lubricants derived from petroleum and another million tons of waste oil from industry are disposed of each year. Most of this disposal of wastes occurs on land, but at least a quarter of the oil from these sources probably reaches the sea via rivers.

The number and the complexity of the sources of crude and refined oil is so great that it is difficult to determine just how much of the liquid hydrocarbon does reach the ocean. Estimates of oil added to the seas vary greatly, ranging as high as ten million tons of oil a year.

Figure 4.13 Leaking pipes and other oil field apparatus are one source of petroleum carried by rivers into the ocean. (Photo by author.)

The Spread of Oil

Slicks and Other Forms: Crude oil is a mixture of as many as thousands of different compounds. Oils from various regions differ markedly in composition and physical properties, and even refined oils are complex mixtures of many types of hydrocarbons Crude oils are soluble in water; some evaporate from the surface, many form extensive and widespread slicks, others settle to the bottom sediments. Depending on which of these processes is predominant, different types of oil spills can have different effects. Full understanding will require extensive future studies.

The physical behavior of an oil slick is poorly understood. Very often, slicks are transformed into irregular, tarry lumps of hardening and breaking caused by wave action and other means. These tar balls regularly foul fine mesh nets in the Mediterranean and North Atlantic and are accumulating in the Sargasso Sea in quantities far greater than the natural masses of seaweed. Thor Heyerdahl describes sailing through long stretches of the Atlantic where the sea was covered by tar balls and clumps of oily refuse.

Oil and Sea Life

All crude oil contains compounds toxic to at least some marine organisms. Nearshore and beach life can be destroyed by being coated with oil brought in by waves and currents just as surface floaters or swimming organisms succumb to the intake of hydrocarbons. For example, the decline of the sturgeon (source of caviar) in the Caspian Sea is probably related to the increasing oil pollution. Hydrocarbons in the form of liquids or lumps ingested by marine organisms may pass through the wall of the digestive tract, become dissolved in the fatty tissues, and held there to be transferred from ingesting organism to predator. This provides a mechanism for passing large quantities of toxic material up the food chain to large fish, birds, and mammals, including man himself.

Attacking Sea Oil: Most hydrocarbons in the sea are eventually decomposed by marine microorganisms. Very little is known about the rates of breakdown, but we do know that no single species of microorganism will break down whole crude oil. Bacteria tend to be highly specific so that several species are usually needed to decompose the various hydrocarbons of a typical crude oil. In addition to this complication, the process of decomposition itself produces by-products that may long endure and may require still other species to continue the process. Ironically, the most readily deteriorated fraction of crude oil, ordinary paraffin, is the least dangerous to life forms whereas the carcinogenic (cancer-producing) aromatic hydrocarbons are broken down very slowly by microbes.

There is some hope that bacteria or other organisms especially destructive to hydrocarbons may be found that will be useful in dealing with the problem of oil pollution. Some success has been achieved in the development of "oil-hungry" strains of microorganisms that have experimentally eliminated small oil slicks in a matter of days. But oil accumulations on an oceanic scale is likely to be quite another matter.

In general, the bacteria and other small organisms that do attack oil accumulations act so slowly that we are probably fast approaching the time when more oil is accumulating than bacteria can eliminate in many regions. In some of the most heavily traveled parts of the North Atlantic, the oil on the surface of the sea has built up into vast films that are moved about by surface current but never fully dissipated.

Other means of treating oil slicks have included burning, which is only possible in certain cases and at the cost of air pollution, and sinking the oil with substances like stearic acid-treated chalk, thereby risking the uncertain effect of oil layers on bottom life. In some instances, very small oil spills have been contained and scooped or sucked up by appropriate machinery.

Detergents have been used in attempts to disperse the Torrey Canyon, and Santa Barbara, and other large oil slicks that reached shore. Some degree of breakdown has been accomplished with large amounts, but the use of detergents produces small oil droplets that are readily ingested by many marine organisms. The combination of detergent chemicals and oil droplets on sea life may be more toxic than the effect of the original oil slick.

Future Outlook

The scale of oil operations is increasing because of the ever-increasing demand for energy (see Chap. 3). The world production of petroleum is nearly two billion tons a year, and 60 percent of it is carried by ship at one time or another. The total capacity of tankers has increased sixfold in 20 years, and the probability of large spills is becoming greater each year with the construction of larger and larger supertankers.

The spills and leakage of oil during recent years is already felt along most sea shores (Fig. 4.14). One stretch of Cape Cod beach contains over 200 gallons of oil per mile. Most Mediterranean beaches have been damaged by oil, and the sea floor there becomes progressively coated and barren of life.

The fact that coastal waters are not devoid of marine life even after decades of contamination by oil spills indicates that the sea is still capable of recovery from this form of pollution. But we do not know how much oil pollution the ocean can accept and still maintain important interchanges of moisture and light between sea and air. The thin film of oil that now persists over great expanses of ocean may already be

Figure 4.14 Oil spilled from passing fuel tanker covers this stretch of Atlantic Ocean near New York City, killing many forms of life and rendering the shore unsuitable for bathing. (Photo by author.)

interfering with the vital processes of algal photosynthesis which supports all marine life.

CROPS AND CATCHES

In Chapter 5 we will consider the problem of meeting the nutritional needs of the world's rapidly rising population in terms of the food-growing potential of our vital soil layer. The ocean is often cited as a vast, still largely unexploited storehouse of food substances that can always be drawn upon to supplement the land and its soil. Let us consider here the actual food we do derive from the sea as well as its potential and limitations for providing tomorrow's food needs.

Present Utilization

Over 50 million tons of food are taken from the ocean each year. Over 90 percent of this consists of swimming fish; the remainder includes shrimp, clams, oysters, lobsters, and seaweed. The amount of food taken from the sea has generally been rising and doubled within the decade of the sixties. Only about one percent of the world's food supply comes from the oceans today, although, it makes up about 10 percent of the world's supply of protein. Unfortunately, there are many indications that rates of extracting food from the sea may not continue to increase.

Sunlit Waters

Over 90 percent of the fundamental organic material on which marine life feeds and depends is produced in the upper, sunlit surface layer of water where photosynthesis of the phytoplankton (floating plants) takes place. These tiny algae serve as the principal food supply for tiny planktonic animals and a variety of fish, also tiny, which in turn, are eaten by other predators (Fig. 4.15).

Some of the phytoplankton and other organic particles from the surface waters settle downward, reaching the bottom to serve as food for the bottom community that includes most of the shelled animals of the sea. Bottom-dwelling (benthonic) animals are most abundant in shallow waters where the organic matter settling down from above travels the shortest distance and accumulates in the greatest quantities. Most of these benthonic animals are filter-feeders, like oysters and clams, that strain particles from the water.

All animal life in the oceans is ultimately dependent on the phytoplankton that float in the sunlit, near-surface waters. Light cannot penetrate into water much deeper than 400 feet, and below 100 feet few plants can survive. Thus, most of the oceans are completely dark, and the fundamental food production is confined to the upper hundred feet or so. Animal life in deeper water, below this photic zone, exists by either moving up to feed, by depending on materials settling downward from these waters, or by feeding on other animals.

The plants and animals of the sea comprise the great marine food chain, many plants being required as food for each of the smallest herbivorous animals and each larger carnivorous requiring a diet of many of the smaller animals to survive. The marine food chain allows the transmission of roughly 10 percent of the energy entering one level above it, meaning that about 90 percent of the energy of the phytoplankton is lost as they serve as food for the tiny animals of the next link of the chain, and so on down the line.

The plankton of the sea have often been pointed to as a source of food for man. We *already* use plankton as food in an indirect sense,

Figure 4.15 The larger fish of the sea depend on smaller animals for food and these on smaller organisms until the phytoplankton, the base of the food chain, is reached. Each pound of a large fish may represent about 50 pounds of the phytoplankton.

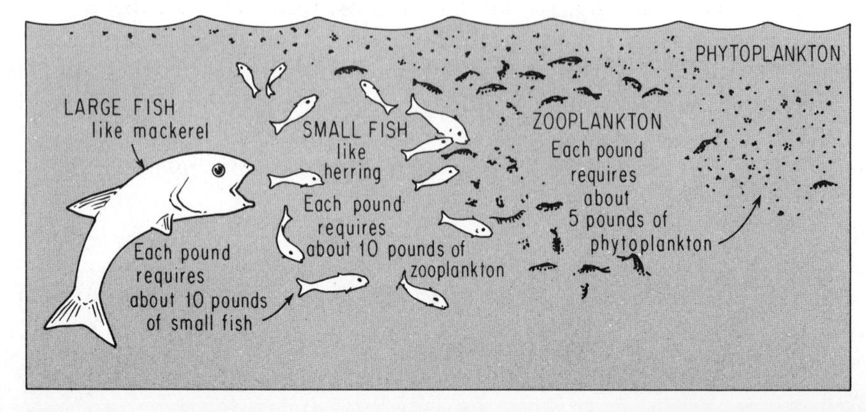

when it is processed through the bodies of the animals that comprise the marine food chain. By catching the larger forms of sea life, we are indirectly harvesting the phytoplankton from the far end of the chain. It is unfortunate that the use of this basic food of the sea must be so indirect. Consider the losses involved when a man eats fish such as mackerel.

One pound of the mackerel required its eating some 10 pounds of smaller herring as food, and at least 100 pounds of tiny, floating animals (zooplankton) had to be eaten by the herring to produce their 10 pounds of flesh. Further, at least 500 pounds of phytoplankton were needed as food for the zooplankton. Thus, it took at least 500 pounds of planktonic plants to yield a single pound of fish consumed by man. With tuna the chain is longer: one pound of tuna fish represents some 5000 pounds of phytoplankton!

By contrast, whales like the baleen or giant blue, that live directly on the zooplankton, are much closer to the base of the food chain than predatory fish like the mackerel and tuna. The baleen whale can engulf entire schools of tiny shrimp-like krill. A single meal for this giant mammal may consist of hundreds of thousands of krill, each of which eats immense numbers of floating plants a day. By taking advantage of the abundant krill, the whale short-circuits the food chain and avoids the need to hunt for other predators of intermediate size.

One pound of the meat of such whales may have required 100 pounds or less of phytoplankton as contrasted with the 5000 pounds represented by each pound of tuna. Since man has yet to construct any device more efficient than a whale to process plankton, he might do well to study ways of cultivating and increasing the numbers of these giants instead of exterminating them by the indiscriminate hunting of recent years.

Finding Fish

Most of the fisheries of the world are located either in the shallow waters of the continental shelves or in certain regions, just outside the shelves, where the water is rising from deeper levels toward the surface and veering off from the shelf edges, causing coastal surface waters to flow outward. These waters carry nutrients causing a periodic enrichment of the surface waters well offshore. Most of these areas of "upwelling" are located off the western coasts of the continents of South America, Africa, and India.

A major example of a fish-concentrating "upwelling" is produced where the Humboldt (Peru) Current flows northward along the west coast of South America (Fig. 4.3B). This major surface flow of water owes its existence to the warm waters that rise just south of the equator. This flow veers southwestward (to the left) toward Australia then further southward and gradually eastward in the latitude of South 50 degrees in response to the earth's rotation and prevailing winds from the west.

Reaching the vicinity of South America, the waters veer again to the left, now northward.

Spreading along the edge of the narrow continental shelf of the continent, the Humboldt Current gradually diverges in a northwesterly direction and causes an "upwelling" and concentration of water-borne nutrients that provide food for the largest concentration of anchovy fish in the world. These swarms are the chief basis for the fishing industry of Peru. Unfortunately, the Humboldt Current varies from year to year in its exact position, and at times the nutrient concentration decreases, and the anchovy crop dwindles as it did in 1972.

In 1965, thanks to the Humboldt Current and the buildup and modernization of its fishing fleet, Peru became the world's leading fishing country. It is ironic for South America that much of the millions of tons of fish caught each year off Peru is ground into fish meal and exported (chiefly to the United States), thereby doing little to diminish the chronic protein deficiencies in a continent that has the world's fastest growing population.

The commercial fishing industry has grown in its efficiency and sophistication, often using search planes and radar to locate schools or concentrations of fish. Such means of intensive fishing have led to a growing danger of overfishing with many species of fish now in drastic decline or near extinction. The herring is one of the chief examples of a species decimated by man's overfishing. Another is the sardine, once of great importance on the Pacific coast of North America. In 1936 and 1937, a fleet of 300 vessels was taking between 100 and 200 tons of sardines a day from the waters stretching from California to British Columbia. By 1955, the annual take had become so large that breeding fell off catastrophically. In three years, the Canadian sardine catch fell from 30,000 tons to less than 500 tons. Sardine numbers shrank drastically, the catastrophe progressing down the coast of Washington and Oregon until, in 1951, the entire San Francisco fishing fleet returned with only 80 tons and the industry closed down in the United States.

Throughout this period of crisis the sardine fleet operators were unregulated and refused to tolerate any limitations that might have allowed the breeding of sardines to continue, thereby maintaining the industry. Whether or not the overfishing of sardines will provide a lesson in conservation for the anchovy industry of South America remains to be seen.

Considering the nature of the oceans, the role of the food chain in limiting the numbers of larger, more desirable fish and the scarcity of highly productive "upwellings" and fish concentrations, it is doubtful that the open sea will provide any vast new food supply for the future. Undoubtedly, the yield can still be increased somewhat. But much more is required than sending out greater numbers of more efficient fishing vessels. In fact, if we double the production of the popular fish caught by presently available fisheries methods, we will probably come close to running out of these species. On the contrary, there is a need to restrict the take from certain species and from areas being endangered by overfishing lest we lose the fish supply we have today.

Many authorities believe that the current 50 to 60 million metric tons of fish taken each year is close to the maximum possible and that this may well be reduced by unregulated overfishing even if the pollution of the sea does not produce the same effect—considering that the amount of man-made waste going into the ocean is expected to increase over five times in the next decade.

Mariculture

Mariculture is the name given to the farming or raising of sea life. Open sea fishing is essentially a hunting procedure fraught with uncertainties and risks of depletion. The development of mariculture is analogous to the shift from hunting wild animals to raising livestock that occurred long ago on the land and would be a logical advance over the present, rather haphazard exploitation of sea life.

Most mariculture projects concentrate on shallow, near-shore parts of the continental shelf that offer the best opportunity to culture both fish and shellfish in large numbers. Bays and estuaries can be especially rich sources of edible marine life with higher potential for annual protein production per square mile than that of the best fishing grounds off any coast. Molluscs, undoubtedly, afford the greatest opportunity for intensive cultivation of sea food in shallow coastal waters. Oysters, mussels, and other benthonic bivalves are filter feeders that live directly on phytoplankton and are, therefore, far closer to the base of the food chain than fish.

The Japanese are foremost among growers of shellfish and have developed advanced procedures for cultivating various species under differing degrees of concentration. In places they grow shellfish on ropes suspended from floats, thereby increasing the numbers of animals possible per unit area and minimizing the problems of siltation and damage by starfish predators, Shrimp are also being cultured intensively along parts of the Japanese coast. Raised from eggs of females caught at sea, the growing shrimp are fed cultured algae, small crustacea, and fish scraps. A variety of food fish have been raised in marine or brackish ponds built and maintained under controlled conditions in Japan and elsewhere.

Pollution and Food from the Sea

Mariculture of shrimp, lobsters, various foodfish, oysters, mussels, clams, and other marine animals, carried out extensively along the coastlines of the world, might theoretically supply all the world's protein needs. But mariculture has been slow in developing, and future expansion is likely to be threatened by growing pollution. Its potential is greatest along coastlines, the very regions where pollution is increasing most rapidly. In many coastal waters, oysters and other shellfish are already being killed by intake of metals and other toxic substances in waste waters discharged into coastal waters. Billions of others of these molluscs are too contaminated by water pollutants to be used for food. For example, the shellfish of Mobile Bay, Alabama, once supplied much of the food for one of the largest concentrations of Indians in North America. The waters of the bay are now closed to oystering and unsafe for shellfishing because of rampant pollution.

In an especially vulnerable situation are the anadromous fish, such as salmon, which migrate from freshwater streams to the oceans during their lifetimes. They must face the multiple hazards of dams, thermal, chemical, and other forms of river pollution as well as the heavy concentrations of pollutants in the estuaries and coastal waters through which they pass on the way to the open sea.

As the amount of heavy metals, pesticides, and other pollutants in the seas increases, the time approaches when fish may simply contain too much toxic substances to be used safely for food for man, or they will dwindle in numbers from the effect of the poisonous substances. Swordfish and tuna caught in some regions have been judged to have too much mercury to be used as food. Many kinds of fish caught off the shores of certain industrialized nations like Sweden may soon have to be eliminated from fishing catches because of high levels of mercury and other dangerous substances.

The building up of toxic metals, pesticides, and oil in the sea, if continued, will do irreversible damage to many organisms vital to the marine food chain, eventually destroying much of the food which we presently obtain from the sea.

Suggested Readings

HEDGPETH, H. W., "The oceans: world sump." *Environment,* Vol. 12, No. 3, 1970.

HOULT, D. P. (ed.), *Oil on the Sea,* Plenum Press, New York and London, 1969.

MARX, WESLEY, *The Frail Ocean,* Ballantine, New York, 1967.

Scientific American, *The Oceans,* (special issue devoted to the oceans), Vol. 221, No. 3, 1969.

STRAHLER, ARTHUR, *The Earth Sciences,* 2nd ed., Harper & Row, 1971.

WILLIAMS, JEROME, *Oceanography,* Little, Brown, Boston, 1962.

5 Soils and Food

Photo courtesy of U.S. Forest Service.

ONE VITAL LAYER

Once, not long ago, our world was large and people few. Now there are nearly four billion humans, and the planet, we suddenly realize, is very small.

These rise in population in recent decades has been at an ever increasing rate, the result of survival rates of infants and children who live to reproduce, giving rise to many more total offspring than would have appeared during the prior years of high child mortality. At present, the world wide average death rate is less than half of the world birth rate, and some countries now are doubling their populations in 20 years or so. It has taken over a million years for mankind to reach our present population (Fig. I.4). Yet in only 35 more years, it seems, this population will be doubled.

Unfortunately, there is no known way indefinitely to increase food supply at a rate corresponding to the spectacular exponential rate in population. Food supply is limited to that which can be grown on the finite land or taken from the finite sea (which has very limited potential; see Chap. 4).

Almost all of man's food supply comes, directly or indirectly, from the soil, the carpet of life that supports almost all of the vegetation on which land animals depend. Soil is the link between the rock sphere of the planet and all living things that grow on its surface. It covers almost all

Figure 5.1 Typical residual soil in road cut, North Carolina. Note the transition from bedrock up into the topsoil that developed by slow weathering. (Photo by author.)

land surfaces, yet is only a few feet or tens of feet deep, astonishingly thin considering its vital role as a foundation for life (Fig. 5.1).

In this chapter we will consider first the nature of the soil, and then its agricultural uses, and ability to provide food for the soaring population.

ORIGIN OF SOIL

Weathering Rock

Wherever a solid rock is in contact with the air, it is being changed. The gases of the atmosphere, its moisture content and temperature, all have an effect on the mineral components of the rock. The result is imperceptibly slow but constant change, the age-old process of *weathering* in which a rock gradually crumbles into bits and particles. This is the manner in which a soil begins (Fig. 5.2).

Many minerals, formed beneath the earth's surface, are unstable when subject to the low pressures and temperatures at the surface and when exposed to the air and moisture of the atmosphere. These minerals change chemically, or break down, into other compounds that are more stable at the earth's surface. Chemical reactions during weathering occur at rates dependent on temperature and on how wet or dry the climate but always measured slow by human standards. During the weathering of a granite, a single grain of feldspar a fraction of an inch in length, may take a century or more to react with the moisture and gases of the atmosphere and be completely changed (hydrated) into the clay, which is the stable end product and a component of the developing soil.

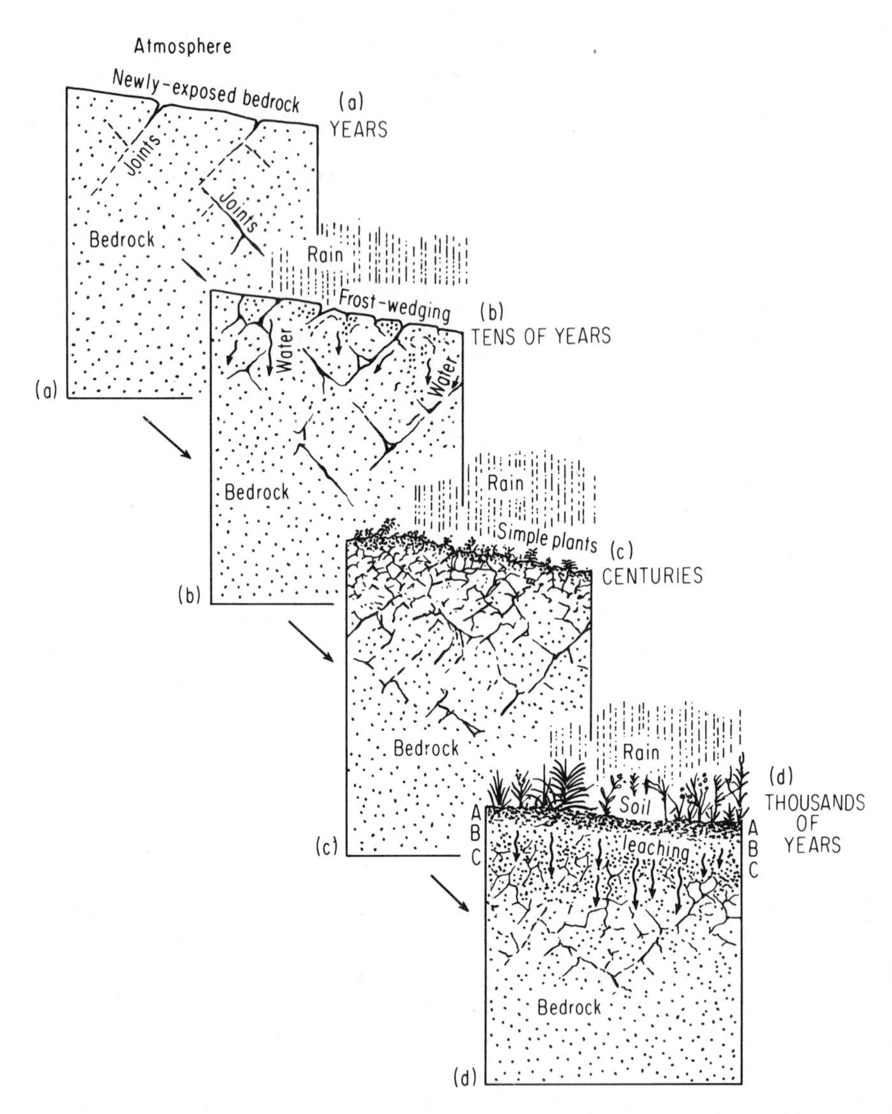

Figure 5.2 Development of soil from initial, bare rock through thousands of years of weathering and mineral breakdown to a well-developed residual soil.

In cold or temperate climates, an important factor in the weathering of rocks is the disintegration produced by the action of frost and ice. When water freezes in cracks in rocks or between minerals, it expands by nine percent and may exert pressure on adjacent surfaces reaching tens of thousands of pounds per square inch. Where enough intersecting fractures in a rock outcrop are wedged apart by frost expansion, pieces of rock may be pried loose from the surface causing the rocks gradually to break down into ever-smaller fragments. Repeated alternations of freeze and thaw are especially effective. Each thaw allows water to seep deeper into cracks to be followed by new wedging and the cumulative disintegration of bedrock into the mineral particles that are the raw material of soils.

Plants and animals also help to break down bedrock into fragments or contribute to the chemical alteration of minerals. Growing plant roots may exert great pressure to enlarge cracks in solid rock and split boulders. Even plants as small as moss and lichen penetrate between mineral grains and loosen the particles of a rock. Chemical alteration of minerals is caused or aided by various organisms including lichen and bacteria.

The role of plants and animals in producing soil usually increases as disintegration of bedrock proceeds and as the coarse fragments become finer particles through which worms and other burrowers can move readily and into which roots can penetrate.

Layering Develops

Soil is a mixture of mineral matter, organic matter, air, and water, in varying proportions. Almost every soil is layered; the different levels or horizons are visible when one digs down below the surface (Fig. 5.3). The part of the soil penetrated by most plant roots, the familiar topsoil, is only a few inches or feet thick. It contains bits of organic matter, usually both plant and animal, living and dead. The quantity of this organic matter largely determines the fertility of the topsoil. Below this lies the *subsoil* that contains little organic matter and that depends for its characteristics on the mineral composition of the source rock. This is usually the bedrock farther below, from which it was derived by the very slow chemical and physical breakdown steps just described.

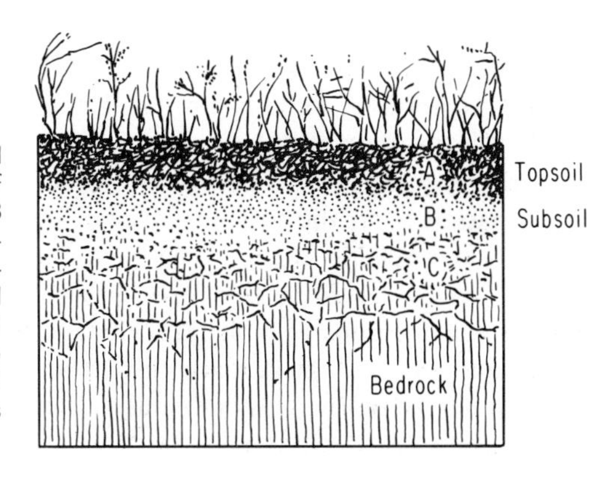

Figure 5.3 A typical residual soil consists of a lower C horizon of broken-up bedrock, an overlying B horizon representing the weathered altered bedrock plus an accumulation of minerals carried from above by descending ground waters, and the A horizon, the vital topsoil, rich in humus and often leached of soluble minerals by ground waters.

Because the nature and composition of a soil varies with depth, it is traditional to distinguish three soil horizons. From the surface downward, these are called the A, B, and C horizons. Of the three, the C horizon is the lowest soil zone. It lies directly on top of, and grades into the bedrock. This horizon consists of partially broken-down rock into which both the original minerals and some new alteration products are present.

In the B horizon or subsoil, weathering has progressed further than in the underlying C horizon. Here the texture is finer and weathering products predominate, with some minerals of the original rock, notably quartz, persisting. In humid climates, weathering products such as clay and iron oxides are concentrated in the B horizon by downward-percolating waters. In arid climates, more soluble minerals like calcite also tend to accumulate, precipitated by water that had previously soaked into the soil and then has been drawn up toward the surface by the high rate of evaporation that prevails in very dry and warm climates. In less arid climates, soluble minerals are normally carried downward through the B horizon by water that ultimately finds its way out of the soil into streams. Together, the B and C horizons comprise the subsoil.

The A horizon or topsoil is the uppermost soil horizon, the one most familiar to us and the layer so vital to agriculture and life. The upper part of the A horizon tends to have a dark color from the concentration of organic material, especially dead plant matter. Into this layer penetrate plant roots and through it burrows worms and other animals. This horizon is sometimes called the zone of leaching because waters descending from the surface have largely leached out or dissolved those products of weathering which are soluble to any degree. Most of the iron oxides have been carried further downward into the subsoil, and some of the fine clay particles have also worked their way downward.

When these three zones are developed distinctly and thickly, the soil is said to mature. Mature soil, passing gradually down through a layer of broken-up rock into the bedrock itself, is found over much of the United States. But in New England, the northern states such as Michigan, and Minnesota, and in most of Canada, the soil is thin, or absent, or ends abruptly against the bedrock. This is the result of the great ice sheets that were present prior to some 10,000 years ago. Wherever glaciers passed over, the soil was scraped off the bedrock and moved elsewhere. As the glaciers wasted back from these areas, bouldery debris was deposited in places from the melting ice. This glacial deposit is sometimes called transported soil. It differs trom the normally developed, residual soil in its clear demarcation from underlying bedrock; its typical difference is in the mineral and chemical composition from that of bedrock and poor development of A, B, and C horizons.

Climates and Soil Types

The dominant processes involved in production of soil from bedrock and its rate of formation will depend on: the nature of the parent rock, climate, action of organisms, and topography. Of these the climate seems most vital. Soils tend to be grossly similar in regions with similar climates in spite of differences in the bedrock from which they were derived.

Climatic factors include the amount of rainfall, humidity, and temperature. These affect the rate of weathering, the breakdown of minerals in the parent rock, and govern the migration of substances within the forming soil, resulting in stratification into the different soil horizons.

Most temperate regions, with their moderate rainfall and cool temperatures, have *podsol* soils. These are characterized by a topsoil rich in humus but relatively thin, with a lower A horizon well-leached to a grayish color and with a heavy, brown-stained clay accumulation in the B horizon (Fig. 5.4). Calcium carbonate and other soluble salts are leached out producing a somewhat acidic composition; this plus their high clay content, make podsols soils of only modest agricultural potential. Gray-brown podsols develop in somewhat warmer climates than podsols; they have been less leached of salts and have a gray-brown color in the lower A horizon. More desirable as agricultural soils than the podsols, they occur widely in the northeastern United States and central Europe.

The agricultural heartland (Great Plains) of the United States, a belt stretching from the Dakotas south through Kansas and Iowa into Texas, is chiefly underlain by two soil types, the *chernozems* and the closely related *prairie soils*. Of the two, the chernozems (called black earth) have a thick, black humus-rich A horizon, originally produced by long-term

Figure 5.4 Four varieties of soil: (a) Podsol; (b) Chernozem; (c) Desert soil; (d) Transported soil of glaciated region.

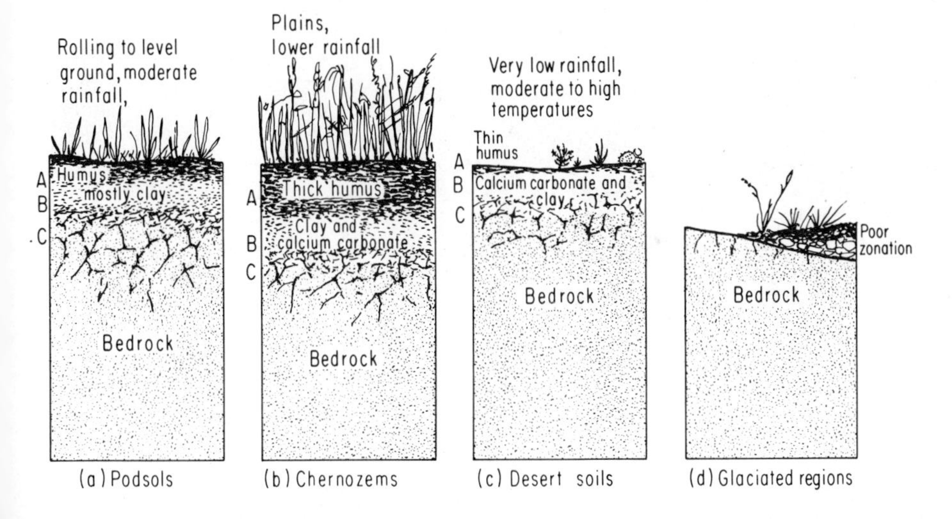

Figure 5.5 (Top) Typical chernozem soil in a wheat field in Kansas. (USDA Soil Conservation Service). (Bottom) Typical desert soil in Utah. (Photo courtesy of Union Pacific Railroad.)

accumulation of decaying grass stems. The leached lower A horizon of the podsols is absent, and nodules of calcium carbonate are usually scattered throughout the B horizon. Chernozems develop in semi-arid, *level* regions where there is just enough rainfall to permit effective chemical weathering, and thick soil formation, and support a thick growth of grass, but not enough to encourage high rates of humus destruction. The lack of steep slopes encouraged soil accumulation rather than erosion. The result was the gradual development of a thick, humus-rich soil, ideal for growing wheat and other grains of modest moisture requirements (Fig. 5.5).

Prairie soil is transitional between the chernozems and podsols. Developed in a somewhat more humid climate than chernozems, prairie soils do not have precipitated calcium carbonate in their B horizon. Otherwise, the two are similar; the thick humus-rich prairie soils have proved ideal for growing corn and many other crops.

Desert soils are unable to support most kinds of vegetation other than sagebrush and cactus plants and are nearly useless for agriculture. As one goes from the semi-arid chernozem soil regions toward the very arid desert soil regions, the thick black soil fades to dark brown (the chestnut soils) and then to the lighter colored brown soils which support only a very spare grass cover and are transitional into the desert regions.

Traveling across the United States along the 40th parallel, from the east coast to Nevada, one crosses a number of major soil types (Fig. 5.6).

Figure 5.6 Generalized soil map of the United States; annual rainfall, temperature, and altitude are principal controlling factors.

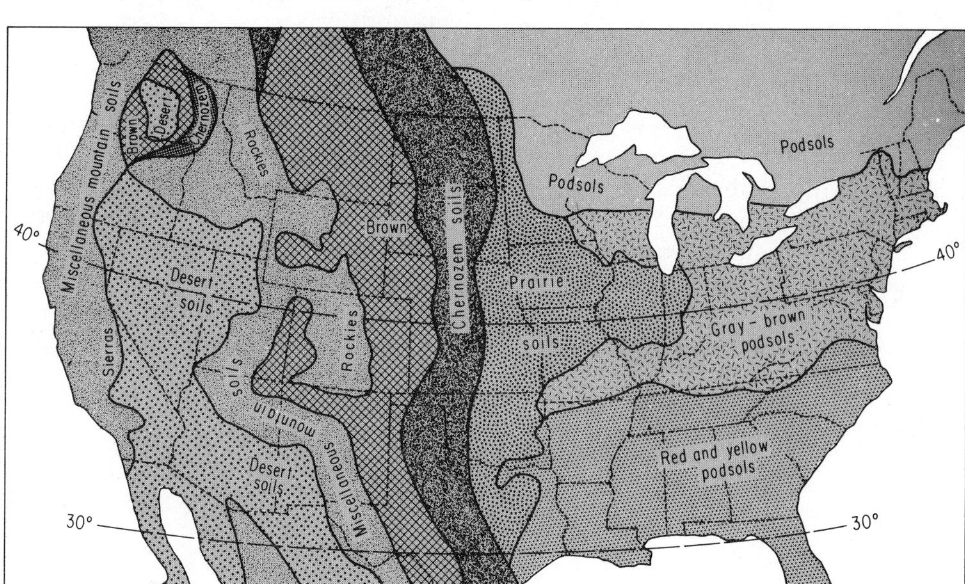

For the eastern third of the country, we travel over the vast gray-brown podsolic soil region. When we have passed west of the Mississippi and Missouri Rivers, we encounter the belt of prairie soils. Appearing next are the chernozems that grade westward into the chestnut and then the brown soils that occur in scattered expanses among the Rocky Mountains where actual mountain slopes support their own variable, high-altitude soils. In Utah and Nevada, we travel mostly over desert soils. Were we to continue to the coast, we would traverse the Sierra Nevada and coastal mountains, again with variable soils and stretches of bare rock.

These east-west differences in soils are chiefly a function of changes in rainfall (see Chap. 1) with generally decreasing rainfall westward. Should we travel in a north-south direction, there would also be a change in soil types encountered, based chiefly on average temperatures. In Eastern North America, we can travel from the red-yellow podsolic soils of the subtropical southeastern states where the high bacterial activity leaves little humus accumulation, to the gray-brown podsolic soils previously described. From there we would encounter the gray podsols of New England and the Maritime provinces and, finally, to the *tundra soils* of northern Canada. These arctic soils are relatively immature because of the slow rate of chemical changes at low temperatures. The average temperature is so low that the deeper soil may remain frozen all year (permafrost), and the upper soil, when thawed, is water saturated and black from organic material that decays only very slowly under the cold, watery conditions.

Outside of the United States, red and yellow podsolic soils may be traced equatorward into lower latitudes where they usually give way to the *lateritic* soils of the humid tropics.

How Long Does it Take?

The formation of soil is dependent on very complex, very slow geologic processes. The rate of production of a good, nutrient-rich topsoil from which animals and man can derive their food depends on the mineral composition of the original bedrock or pre-soil material, the climate, and the abundance of plant life which develops. On newly formed or recently exposed bedrock, a topsoil thick enough to support crops will usually take *thousands* of years to form (Figs. 5.2 and 5.7). If we start with soft, alluvial sediment, with loose sands or volcanic ash, centuries will probably be required for a useful topsoil to develop. Under some conditions, topsoil may be encouraged to form more rapidly (see below) but in terms of a human lifetime, soil is *not* a renewable resource.

The development of a life-supporting new soil on presently barren ground by natural processes is so slow that it can be of little practical

Figure 5.7 The formation of soil is extremely slow by human standards. This surface of lava that cooled about 500 years ago has still not developed a soil in spite of the centuries of exposure to atmosphere and rain. Craters of the Moon, Idaho. (Photo by author.)

importance or of serious consideration as a factor in the present world food supply crisis. Once a topsoil is totally washed away or otherwise removed, the region is lost to agricultural use within the lifetime of all living persons and of several succeeding generations.

GROWING FOOD

Early Soils

Rocks have been weathering for billions of years, for at least as long as there has been an atmosphere. But it was not until about 400 million years ago, when living things first spread up from the sea that organic matter has been available to mix with weathered rock debris and to build up our essential, life-supporting topsoils. Virgin soil owes much of its fertility to the billions of generations of plants that have grown, lived, and died, each year adding a bit more organic matter to the topsoil.

Organic matter in a soil usually has two forms: first, the plant and animal remains, still undecomposed or in process of decomposition and second, the humus, that fraction which is well-broken down into the colloidal state in which it is usable to new plant growth.

Humus is valuable basically because of its abilities to absorb and hold soil water available for plant life, to filter dissolved nutrients from soil water and return them, even after the water has dried, to be used for plant growth when water reappears. In addition, the colloidal character of humus is very important in holding the soil particles together. Most humus contains a variety of useful substances including carbohydrates, chiefly of plant origin and proteins of animal and plant origin.

Long-cultivated soils usually differ greatly from the original or virgin soils of the same region. The original, uncultivated soils can be said to have been in equilibrium between plant growth and the nutrients of the soil. Plants that are supported by these soils are those which are well-suited to the various nutrient substances found in the soils. When such a soil is plowed and farmed, a different vegetation, usually a single species in great concentration, is imposed on it (monoculture) and the natural balance is lost.

When a virgin or long-uncultivated soil is put under the plow and overturned, the higher degree of aeration begins to alter its composition. Breakdown of organic matter becomes much more rapid; nitrifying bacteria increase their activity, but the nitrates formed are washed out rapidly by the increased drainage from the fields. Harvesting removes much of the new plant growth, leaving only roots and stubble to be returned to the soil; thereby, more and more of the vital nitrogen is lost with each harvest.

During the century or two that intensive farming has been going on in the United States, vast expanses of rich, virgin soil that required hundreds of thousands of years to build up have rapidly deteriorated. The first step taken by pioneer farmers was often to burn off or remove brush and other native vegetation and, with it, much humus and potential humus. Crop residues were often burned to simplify the next plowing. Plowing without regard to the slope of terrain was normal practice, usually followed by the erosion of topsoil (see losing ground below).

Increasingly powerful farm machinery has been repeatedly stirring up and turning over the soil in most farm regions for more than half a century. Gradually, the basic structure of soil and subsoil has been disrupted, and more clays are brought to the surface until soils absorb water less effectively and their natural nutrients diminish. In most regions, the combination of monoculture and other expedient but short-sighted farming practices have reduced the natural fertility of the soil to less than half that of the original land.

Plant Nutrients

The amount of crop plants that can be supported on a land surface depends, in large part, on the nutrients available to them. For example, carbon is obtained by assimilation through leaves of the carbon dioxide in the atmosphere during photosynthesis. Most of the oxygen is obtained from soil water through the roots; some is taken from the air. Hydrogen is also obtained from soil water. Nitrogen originally is produced in the soil through bacterial action but is chiefly taken into plants from dead organic matter.

Nitrogen, phosphorous, and potassium are the three indispensable elements that control soil fertility. All three are usually present in soils but, in the soluble state required by plants, they are produced rather slowly. Because growing plants absorb nitrogen, phosphorous, and potassium in considerable quantities, these are the elements that may be rapidly exhausted in a soil. It is for this reason that yearly removal of harvested crops can result in a serious loss of the three critical elements and may require their replacement by adding fertilizer.

INCREASING THE YIELD

One hundred years ago about 95 percent of the American population was engaged in agriculture. At that point in our history, the land grant colleges were instituted, and a new concentration on agricultural research and education began that was to change drastically the capacity of the land to produce food for the growing population. Year by year, an increasing percentage of the population has concentrated in urban regions, depending on the industrialization for livelihoods while an ever smaller percentage of the population concentrated on growing the country's food. Although new western lands were added to the total agricultural acreage of the United States, the principal factor in enabling this small number of farmers to grow food for the great remainder of the century was the rapid improvement in efficiency of agricultural techniques that increased the crop yield per acre. These techniques include mechanization, chemical fertilizers, pesticides, and the development of new plant varieties.

The increasing yield per acre has been especially spectacular in the two decades starting in about 1940, increasing more slowly thereafter. Today less than three percent of the American population does the farming and provides the food for the remaining population.

Fertilizers

Intensive agriculture tends to deplete the soil of certain natural nutrient elements that must be replaced if good crops are to continue. We have considered the various nutrients needed and extracted by plants from their soil environment. To the plant it does not matter whether nitrogen and other vital elements originated by the decomposition of earlier vegetation or was added to the soil in the form of organic matter

Figure 5.8 Chemical fertilizers are being added to farmland throughout the United States. (USDA Soil Conservation Service.)

like animal manure or as inorganic chemicals. This fact is the basis for the use of modern chemical fertilizers.

On most large farms, it has proved practical to replace organic fertilizers with chemical additives of inorganic origin (Fig. 5.8). Large quantities of chemical fertilizers have been used in the United States, Japan, and in other regions of high crop yields per acre of farmland. Before considering the advantages and disadvantages of these modern inorganic fertilizers, we will consider the more traditional, organic fertilizers available to agriculture.

Manure: Green manure is fresh plant matter that is plowed into soil in order to increase the content of organic matter and humus available to the next crop of vegetation. It may be simply the weed and random growth that have taken over a field left uncultivated for some time, or the grasses and clover that survives from a field previously used for pasture, or it may consist of a crop intentionally grown to be used as fertilizer. Plowing under of plants at an ideal growth stage may add as much as 100 lb. of nitrogen per acre.

Farm Manure: Farm manure produced by cattle and other livestock can be an important element in maintaining fertility of farmland, just as the droppings of browsing and grazing wild animals have always been a natural way of returning nutrients to forests and grasslands.

It was once common practice to turn cattle loose on the stubble and grass left after harvest, thus adding manure to the soil. For centuries, horses drawing plows and other machinery have been leaving manure in the soil they help till. But the replacement of horses by motors on most farms and the recent tendency toward a concentration of cattle raising in certain localities has meant that manure would have to be transported from the stock-raising farms to the crop-growing farms if it is to be used as a fertilizer. The transportation and labor costs are usually prohibitive. The rapid decrease in the number of mixed farming operations (crops and cattle) in the United States has been one reason for the increase in use of chemical fertilizers.

Compost: Compost is a man-made mixture of natural ingredients used as a fertilizer. It is usually produced by piling up waste organic matter in layers on which slow processes of decomposition act to produce a rich humus. Well-prepared compost is richer in plant nutrients than most farmyard manures, partly because of the time involved in its formation during which oxidation and other changes reduce the bulk and concentrate the nitrogen and other nutrient substances.

Composting usually involves the building up of alternating layers of carbohydrate-rich plant waste (cuttings, leaves, vegetable refuse, etc.), a protein-rich animal waste (manure, animal remains), and topsoil to provide the microorganisms that aid the decomposition processes.

Compost can be a valuable substance for enriching soils, but it is so limited by the materials, labor, and space available for its production that it is rarely used on large farms.

Sewage as Fertilizer: There has been an increasing amount of speculation concerning the use of urban sewage as fertilizer. Asia and other regions have long used human waste to enrich soils and increase farm yields, but in the United States modern sewer systems ensure that none of this waste is returned to the land. The sewage of major urban regions includes high amounts of nitrogen, phosphorous, and potassium, which originated in the soil and the growing plants on which man ultimately depends for food. Many environmental scientists have been studying methods of recycling and salvaging the elements in at least some of this sewage.

One of the chief difficulties in using urban wastes to fertilize the soils of farm regions is the separation of useful from nonuseful (even poisonous) substances in the general urban sewage that is a complex mixture of organic matter and chemicals of various kinds. These problems have been overcome in some parts of Europe where farms using urban sewage as fertilizer are operated successfully, but the odors involved have required that such farms be at a considerable distance from the city source.

It seems inevitable that future food growing requirements will greatly increase the amount of world farm acreage fertilized by urban sewage.

Chemical Fertilizers: The rapid growth of chemical industry in recent decades has made available a variety of artificial fertilizers that are cheaper and more easily usable than organic fertilizers. In the last quarter of a century, the use of nitrogen fertilizers alone has increased over seven times and dwarfed the use of organic fertilizers. In some regions where soil has been depleted of natural nutrients, the results of adding nitrogen fertilizer have been spectacular, with annual yields more than doubled.

Unfortunately, the use of nitrogen fertilizer has its limitations and even its disadvantages. After decades of use, it has become increasingly evident that an application of the same amount of nitrogen-containing fertilizer to soils each year does not produce crop yields of similar size, year after year. In most cases, the yield will eventually drop off unless an additional amount of fertilizer is used. This phenomenon indicates that the *natural* ability of the soil to produce usable nitrogen is decreased by the presence of nitrogen in chemical fertilizers.

The broad valleys of California have been the site of chemical fertilizer application on a vast scale for many years. The steady increase in the quantity of these fertilizers used in California valley farms is partly the result of the necessity to add more nitrogen fertilizer each year to compensate for the diminishing ability of the soil to produce its own nitrogen.

Growth of plants by using large quantities of nitrogen fertilizer seems to be a chemical process that diverges more and more from the natural growth process. There are indications that we are reaching limits to the yield increases that may be brought about by additions of chemical fertilizers. In parts of India, increased use of nitrogenous fertilizer has recently doubled the yield of rice per acre. But in Japan, where great quantities of the chemical have been made, and used, and where rice production had risen significantly after World War II, there has been a leveling off of yields since about 1960.

Nitrogen and phosphates are now being added to American farmlands in such concentrations that excesses find their way into groundwater, brooks, and rivers, increasing the chemical content of our water bodies. Nitrates and phosphates building up as pollutants in lakes have been causing vast algal blooms with decreased water circulation and stagnation, followed by decay of the masses of algae and resultant oxygen depletion of the water. This diminishes the rates of sewage purification in rivers by the natural aquatic processes which require oxygen (see Chap. 2). Thus, after a few decades, agriculture's annual gain in productivity by use of chemical fertilizers is reaching limits while the fertilizer by-products in the form of water pollutants (added to by domestic detergent chemicals) are threatening the waterways and the health of the country.

Pesticide Use: The use of insect and other pest-killing chemicals to prevent crop damage or failure and to increase yields has been highly developed in the United States, especially since World War II and the discovery of DDT. An arsenal of chemical weapons, chiefly chlorinated hydrocarbons, has been developed and are annually sprayed on American crops. Crop dusting to kill plant damaging pests and to increase agricultural yields has become a common part of the American scene, and the use of pesticides is considered vital by many agriculturists and world food planners.

But insecticides may act as poisons to some of the microorganisms and to other plant and animal life within the soil that are so vital to its fertility, for example, the burrowing earthworms. The population of these and other kinds of soil fauna and flora tend to be decreased by the buildup of insecticide residues. Studies have demonstrated that even years after farm acreage has been sprayed with such insecticides, a large percentage still remains as a potentially dangerous constituent of the soils.

The opposition to the use of pesticides has been growing rapidly since the discovery of residues of DDT in rivers, lakes, and in ocean waters (see Chap. 4).

The use of DDT has been limited by law in the United States, but it is still used here for some purposes and used very extensively in many other parts of the world. Other chlorinated hydrocarbons, some safer than DDT but some, like dieldrin several times as deadly, are widely used for agricultural purposes. The DDT content of our own bodies has been increasing for years, and potential danger to life posed by pesticide buildups may be the deciding factor in limiting or even halting the use of agricultural poisons for destruction of agricultural pests.

Research stimulated by growing awareness of the dangers of chemical pesticides has developed some promising biological methods of pest prevention. Concentrating certain birds or other natural enemies has been successful against some crop-damaging insect species, and various genetic approaches are being developed to decrease the effctiveness of pest reproduction.

Irrigation

Water is a vital plant nutrient; without it no crops will grow. Irrigation has long been a means of adding to the water available on dry farmlands. Waters redirected by pipes or ditches from rivers, distant mountain streams, or deep wells are used to supplement the water available for plant growth in regions of limited rainfall, thereby increasing the crop yields (Figs. 5.9 and 5.10).

In most arid regions of the world, some form of irrigation has been used for thousands of years. Indians in Arizona and adjacent regions

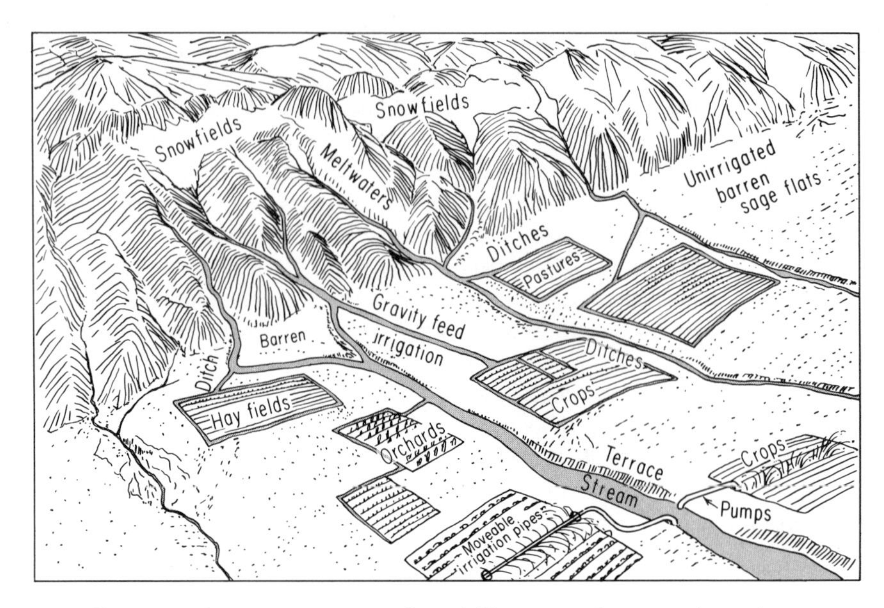

Figure 5.9 Irrigation requires a dependable source of water and may be distributed by pipes, canals, and ditches to fields and orchards. Most irrigation depends on gravity flow from rivers that rise in mountainous regions. Pumps are used where fields are immediately above the river used as a source.

Figure 5.10 Irrigation canal in Utah. Source of water is from a river fed from snows visible on distant mountains. (USDA Soil Conservation Service.)

had developed irrigation ditch systems long before the Spaniards arrived. The first large-scale irrigation in the United States by Europeans was that undertaken by Mormons in Utah in 1847. Ten years later, a similar irrigation development was begun in southern California by German immigrants. Since that time many states, especially in the Southwest, have developed intensive irrigation. These arid regions are usually sunny and warm, with long growing seasons, so that rapid growth and multiple harvests are not unusual with modern methods of irrigation.

There is no question of the value of irrigation in increasing crop yields but in most presently nonirrigated dry lands, there are many barriers to future irrigation projects, and many presently irrigated regions face serious problems in the future.

Major limitations to extension of irrigated acreage include the cost and availability of water. In addition, very arid regions are often so low in soil humus because of previous sparse plant growth that major fertilization projects would have to be carried out, in addition to irrigation, in order to grow many of the most needed kinds of crops.

Irrigation of dry lands with their high evaporation rates requires water in enormous amounts (see Chap. 2). With all of the ingenuity of science and government at work and in spite of all the great dams and reservoirs constructed in the past half century in our vast Southwest, that region is still very far from being covered by lush vegetation. Most of the reservoirs are filling with sediment; the water is evaporating rapidly in the hot sun and low humidity. Communities downstream from dams have often suffered rather than gained while relatively little increase in agricultural productivity has taken place.

One of the world's best known examples of a dam intended to increase a nation's agricultural capacity is the new Aswan Dam in Egypt. Finished after a decade of construction, it will form a reservoir which is expected to increase the arable land of Egypt by $\frac{1}{3}$ by making abundant water available for irrigation, and, thereby, greatly increasing the food production of the country. Unfortunately, the dam will bring to an end the annual flooding of the Nile during which organic matter-rich silt was washed down from upstream regions, a natural process of soil enrichment that has provided the agricultural basis for life in Egypt for thousands of years.

This is one price to be paid for the increased irrigation possibilities and the electric power to be provided by the dam. It is hoped by the government planners that the gradual loss in soil fertility caused by the new barrier will be compensated for by application of chemical fertilizers largely produced by use of the cheap power expected to be available at Aswan. Unfortunately, the greater harvests possible in Egypt will be more than consumed by the populace that almost doubled during the construction of the great dam.

Complicating the Aswan picture is the recent recognition that cutting off the nutrient-rich waters of the Nile has caused a sharp drop in the numbers of fish and shrimp taken from the Nile delta waters and from the Mediterranean in the delta vicinity. This marine life has long served as food for the Egyptians as well as for other Mediterranean peoples.

Desalinization: Nearly all of the land that can be readily irrigated with river water is already being irrigated. With few rivers left to be tapped, the further irrigation of any major part of the vast expanse of arid wastes in the world requires the use of the sea itself. This requires two major accomplishments:

1. The removal of salt from seawater, *desalinization*
2. The distribution of the desalted water within the dry lands

Desalinization of seawater has become increasingly important as a supply source of drinking water in arid regions, like Israel, Libya, and Egypt. Desalinization has been accomplished commercially by the well-established distillation methods and by newer freezing techniques. Both of these procedures require the availability of large amounts of energy that is usually produced by burning quantities of coal, oil, or gas (see Chap. 3). At present, crops grown by use of water obtained by desalinization would be very expensive unless the cost of farm and technical labor was unusually low.

To feed a populace entirely on crops (or animals living on crops) grown entirely by desalinized water would require, yearly, the energy produced by burning hundreds of gallons of oil (or its equivalent) for each person. But even this great expenditure of energy applies only to regions located very near the ocean. For inland regions, still more energy would be needed to transport the water to the irrigated fields. For most inland regions, the cost of major engineering works, such as tunnels through mountain ranges, would have to be added to the already expensive water.

Overirrigation: Saline or alkali soils are poor soils that have been affected by the addition of salts, such as sodium chloride and calcium carbonate. Under humid conditions these substances are usually carried down by downward-soaking rain waters into the water table, to seep out into streams and rivers. But in arid regions, this downward removal of salts by water, called leaching, is usually not as effective, and the concentrations of salts within lower soil levels is common.

Figure 5.11 Regular irrigation in dry climates, like in New Mexico, may cause buildup of salts in upper soil, thereby rendering fields unsuitable for crops. (USDA Soil Conservation Service.)

In extremely arid regions, salts tend to be drawn upward by the evaporation of rainwater which had first soaked down and dissolved some of the soil salts. The result is a concentration of salts in the uppermost soil (Fig. 5.11). A situation of this sort may be artificially produced where soils are heavily irrigated in arid regions. The rising of the water table during irrigation together with the high evaporation rate bring salts up into the topsoil where they are harmful to plant roots.

As a remedy, water additional to that needed by the crops may be periodically added and allowed to soak down into the soil for long periods of time in order to leach away the salts from the topsoil. Such countermeasures are complex, costly, and often only partially effective in the irrigated regions.

A prime example of overirrigation is the vast Imperial Valley of California, one of the most valuable agricultural regions in the world. The sunny climate permits intensive cultivation, and the Sierra Nevada to the east gives rise to the river waters which are directed into an intricate system of irrigation ditches and allow multiple, rich crop cycles. But this intensive irrigation has finally produced widespread, damaging soil salinization. Each year there is an increase in the number of salt-whitened fields in which plant growth is stunted or wilted.

In Pakistan, where over 40 million of the population depend on some 25 million irrigated areas, the elevation of the water table by heavy irrigation has carried salt to the surface and has made so many fields saline and nonproductive that for several years there has been an annual loss of about 100,000 acres of cropland.

LOSING GROUND

Soil is the most valuable part of the solid earth, the thin layer that supports man and all other land life by the plant growth that it nurtures. Not only food but also most of our clothing is provided directly or indirectly by this critical layer. As the gases and moisture of the air, acting together with hosts of tiny plants and animals, slowly disintegrate solid rock to build up a fertile soil, the layer of soil is acted upon by various natural erosional processes, and some of it will be washed or blown away with time. This natural erosion is a very slow process which has been going on throughout geologic history, one that normally is balanced by the slow decomposition of rock which produces more soil.

Accelerated Erosion

Today there has been superimposed on this dynamic cycle of soil formation and natural erosion the dangerous process of *accelerated* erosion caused by man's destruction of protective plant cover, and the overuse and poor management of soils (Fig. 5.12). In the United States, millions

Figure 5.12 Soil erosion on sloping farmland. Much of the topsoil has been washed away. (USDA Soil Conservation Service.)

Figure 5.13 (Top) Clear cutting of forests usually accelerates soil erosion because of the destruction of tree roots and protective humus accumulation by logging. (Bottom) Aftermath of destructive logging: topsoil has begun to wash downslope; siltation visible.

of acres of good cropland are lost annually to agriculture by erosion of their topsoil; this amounts to over $2\frac{1}{2}$ billion tons of soil washed or blown from our farmlands each year. More than a quarter of this is carried off to the sea by the Mississippi River. Most of the remainder also reaches the oceans, carried as sediment by the other rivers of the country.

Soil erosion is the deadly enemy of agricultural productivity. The rate of soil erosion is always increased wherever forests and other original vegetation has been cut away. Wherever networks of live tree roots no longer exist, the runoff of rainwater is greatly increased. After a heavy rain, the unhindered waters seep out at the surface in myriad springs; trickles of water merge into brooks on the barren slopes and race downhill to erode gullies in any meadows or farmland located at lower elevations.

After a region is timbered, there is usually a great increase in the amount of sediment washed by rain into streams (Fig. 5.13). This mud and silt, once part of the woodland soil, is carried down into bottom lands where it may clog irrigation ditches and culverts and, otherwise, interfere with a planned agricultural region. Such sediment may bury fertile acreage beneath a blanket of coarse, alluvial material unfit for crops.

Figure 5.14 Soil erosion from poor farming practices. (USDA Soil Conservation Service.)

When the soil is disturbed by plowing and plant cover is periodically absent during agricultural cycles, the soil is more susceptible to erosion by rainwaters and wind. It usually takes 10 to 15 years of cultivation for virgin soil to begin to show signs of erosion. After that time, even gentle slopes begin to lose topsoil from erosion by sheets of rain while gullies start to form on steeper slopes. A quarter of the cultivated cropland in the United States is now being damaged by erosion at a critically rapid rate (Fig. 5.14). Another quarter is being eroded at a slower but measurable rate. Nor is much of the remaining farmland free from some measure of erosion.

Overused, Thinning Soils

Cultivation Losses: The loss of topsoil from cultivated fields varies widely from one region to another. In 1939, the cornfields of Wisconsin lost 80 tons of soil per acre, which means that all of the topsoil (average: seven inches thick) would have been removed in 11 years if erosion had continued at that high rate. Farmlands in hilly states like Maine currently suffer rapid erosion by rainwash. Annual soil erosion rates in parts of Maine have been measured at an average of 18 tons of soil lost per acre. In western Iowa, steep-sloped farms have had losses of over 30 tons of soil per acre in some years. The situation has now been greatly improved by better plowing and planting techniques that have lowered the annual erosion rate in places to less than five tons lost per acre.

The drier the climate, the more critical is the role of plant life in preventing the slowly-formed soil from washing or blowing away. In the dry, western grasslands of the United States, prior to agriculture, the delicate soil structure was topped by stabilizing grass and roots. But when farming began, plowing disrupted the soil structure; the capacity of the soil to hold water was decreased; crop planting left the land bare for part of the year. Thus, the scene was set for severe erosion. On the southern Great Plains, the droughts of 1890 and 1910 brought seasons of poor crops and accelerated erosion. In 1931, a particularly severe drought occurred in Oklahoma, Kansas, and northern Texas. The native vegetation was destroyed and the soil structure disrupted; the dry plowed lands were swept by strong winds, and the topsoil and upper subsoil were carried off (Fig. 5.15). Some of the former soil piled up as dunes, the finer components were blown thousands of miles eastward. This was the Dust Bowl, one of the miseries of the depression years that led to the migrations of farmers, described in Steinbeck's "Grapes of Wrath." In the 1950s, droughts again resulted in crop failures and accelerated erosion in the Great Plains. Much of this region can no longer be used for

Figure 5.15 "Dust Bowl," Oklahoma in 1937. Disasterous wind erosion of topsoil has destroyed this farmland that was too dry for the kind of cultivation being practiced. (USDA Soil Conservation Service.)

farming. Thousands of these dry acres should never have been planted in the first place.

Overgrazing Losses: Another major cause of soil loss is heavy grazing by cattle and especially sheep (pulling out plants roots and all). Overgrazing causes exposure of increasing areas of bare ground and leads to extensive erosion by rainwash and wind. As the result of animals grazing in excess of the ability of the lands to support them, and the water and wind erosion which followed, many parts of our western grasslands and plains are now nearly bare of grass (Fig. 5.16). Great expanses of South Africa have also been overgrazed and transformed from grasslands into scrubby, half-bare plains.

Devastated Lands: The eastern Mediterranean countries provide many examples of man's destruction of the vital soil layer through overuse. In hilly regions of Syria and Lebanon, most of the topsoil has been washed away, and acreage which once supplied food enough for thousands of persons can barely provide food enough for hundreds now. Primary cause of the soil loss in these relatively dry regions is the cereal-grain farming which is much less effective in protecting the soil than was the original vegetation.

The demand for wood for fuel and construction led to decimation of the famous, original forests of Lebanon. Once cut over, the former wooded areas became grazing ground for sheep, and barren stretches expanded steadily as the topsoil washed or blown away from the lands no longer protected by networks of roots and overgrazed by the animals.

(a)

(b)

(c)

Figure 5.16 (a) Cattle on overgrazed range in Idaho. (U.S. Forest Service.) (b) Overgrazed field on right contrasts with normal grass on left. (U.S. Forest Service.) (c) Extreme overgrazing has created near-desert conditions in this part of Wyoming. (U.S. Forest Service.)

Empty slopes now replace the great forests of Lebanon, the life-supporting topsoil worn thin or gone completely.

To the south of the Mediterranean, the Sahara Desert has been widening as peripheral, populated regions are overgrazed and overused. Far to the east, broad expanses of western India, once covered by trees, are now desert.

Man has now destroyed half of the world's forests and almost doubled the amount of its desert lands.

Nonagricultural Erosion

Man causes erosion and loss of topsoil in many ways other than by poor agricultural methods, overgrazing, and timbering. Some of the most striking examples of accelerated erosion are to be seen at building construction and road building sites (Fig. 5.17). During heavy rainfall, the soil cover of construction-scarred land may be washed away at a rate ten times that which occurs on a plowed field of similar slope. A quarter of the sediment in the Potomac River in recent years has been washed from construction sites around Washington. The erosion is increasing each year as the clearing of land for construction purposes continues to keep pace with the growing population.

In West Virginia, Pennsylvania, Ohio, and nearby states, the practice of strip mining of coal is totally eliminating the value of large parts of agricultural and forest regions even as it alters and destroys the landscape (Figs. 3.10 and 3.11). In order to reach buried coal seams from the top of a hill, bedrock and soil together are usually bulldozed off and dumped into the nearby lowland, thereby eliminating the crop potential of the buried lands. Strip mining also means the cutting down of trees and removal of vegetation cover, thus, increasing the surface runoff after rains and immensely increasing the rate of erosion in a region.

Soil Conservation

Cultivation of the land invariably involves the erosion of topsoil, but the loss can be minimized by adherence to some basic techniques of conservation farming. One of the best known and most important of these techniques is *contour plowing* (Fig. 5.18). This consists of plowing curved furrows at right angles to the slope of the ground. This practice produces horizontal crop rows rather than rows which run up and down hill or oblique to the slopes. The result can be a significant decrease in the rate of surface runoff after a rain thereby minimizing the amount of erosion caused by the rainfall (Fig. 5.19). Each ridge of earth raised by contour plowing for crops serves to hold back or impound the rainwater so as to allow slow percolation of water to plant roots long after the rain has ceased.

Figure 5.17 Clearing woodland and farmland to build industrial or residential development results in loss of soil by erosion and loss of land for further agriculture. (USDA Soil Conservation Service.)

Figure 5.18 Contour plowing is a soil conservation technique of great importance on sloping ground. (USDA Soil Conservation Service.)

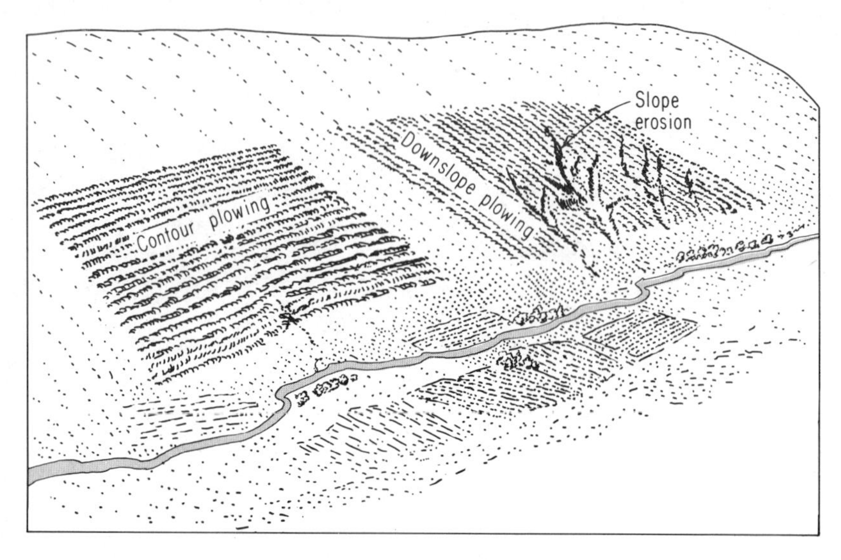

Figure 5.19 Contour plowing versus downslope plowing and resultant erosion. Erosional gullies in the field on the right begin when rainwater channels down plowed furrows.

Strip cropping is a soil conservation technique that consists of planting strips of close-growing plants, such as grass or clover, in alternate strips between other strips of the more valuable crops that consist of plants growing relatively far apart. The strips of close-growing plants slow the runoff after a rain and decrease the overall rate of erosion of the field.

Crop rotation can also be a means of slowing erosion where a periodically grown secondary plant crop consists of a closer-growing variety that allows less soil erosion. *Cover crops* are dense growing plants used to protect fields at a time when there is no other crop being grown, and the temporarily barren soil would, otherwise, be subject to erosion by wind and water.

To deal with the severe erosional problems in many parts of the country, the United States Government has established numerous Soil Conservation Districts in which reforestation of cutover and overgrazed slopes takes place in order to minimize erosion and enlightened conservation farming practices are encouraged. Such practices include contour plowing, or terracing, crop rotation, suiting the vegetation planted to the soil type and climate, and redeveloping some croplands into grasslands for grazing.

In spite of conservation farming, about 400,000 acres of American land are lost each year to agricultural use by erosion—this at a time when food supply is more critical than ever before.

THE UNUSED LANDS

The most optimistic estimates by agricultural scientists state that only about $\frac{1}{4}$ of the total glacier-free land of the world (about 7.8 billion acres) can *ever* be used to grow crops. But considerably less than half of this potentially arable acreage *is* being used for agriculture. Does this mean that billions of acres remain to be planted and harvested? The concept of "potentially arable" is misleading. Almost all of the land that can be cultivated under local economic situations is now being farmed. Millions of acres are essentially swamplands, and other millions would require extensive and costly irrigation (see previous heading). Extensive tracts of potential farmlands are beyond the reach of any existing roads and are often heavily wooded enough to require monumental clearing efforts.

Jungle Lands

The equatorial jungle areas have often been described by food planners as among the principal reserves of essentially unused lands which could be cultivated when we run out of fertile land elsewhere. In Brazil, jungle soils have long been cultivated in a primitive way by burning and clearing isolated strips for a few seasons at most, after which they are allowed to grow back into jungle. But recent pilot attempts to farm Brazilian and other tropical regions in a regular, intensive, mechanized manner have led to sobering realizations.

Under a thin top layer, the tropical soils are usually laterites, soils with high concentrations of aluminum and iron and from which most silica and other substances have been leached out by the intense downward-soaking rainwaters. Once the jungle has been cleared and the thin, dark topsoil is dispersed, and as the wet soil is plowed and exposed to the sun, the iron-rich clays of the soil begin to dry up. After several seasons of repeated plowing and exposure to the sun, the lateritic soils are baked into brick-like hardness and are essentially useless for planting crops. In fact, the laterite soils are often more useful for building than agriculture, as witnessed by the durability of the ancient Cambodian city of Angkor Wat which was built chiefly of blocks of laterite. Abandoned centuries ago, it still stands as a fantastic architectural complex in the jungle.

Swamplands

The conversion of swamps into croplands by draining procedures has been going on for many centuries. Swamps and marshes can be thought of as flat areas in which the water table is coincident with the ground surface. They are usually formed by the filling in by sediments of a lake or pond that was itself caused by the blockage of a stream or river by rock debris or beaver-built dams.

Few plants will grow in water or with their roots in ground that is constantly saturated. But swampland may be high in organic matter because of incomplete decay of plant substances deposited there and can become rich, level farmland if the water table can be lowered.

The water table in swampland can often be lowered by digging ditches that provide avenues of surface water flow, allowing ground water to drain more effectively. With a lowered water table, the entire swamp may become available as a meadow or field of agriculture. Drainage of swamps as a means of increasing acreage available for agriculture is a method restricted largely to the humid regions where swamps are likely.

Large scale drainage of swamps has occurred in Europe to a greater extent than in North America. The fertile Fen district in England was once extensive swampland. In Holland, salt marshes once flooded by the sea at high tide, have been transformed into fields for growing crops.

The practical gain of draining a swamp, in terms of the new land available for agriculture, should be weighed against the irretrievable loss of the swamp as a component of the landscape with a unique combination of flora, fauna, and landscape. By converting the swamp, man may have added a piece of farmland by destroying a scenic treasure, never again to be used as a refuge for bird or animal.

Old Farmlands

During the 12 years ending in 1970 there was a *decrease* of over 10 million acres of land being used for farming in the United States (Fig. 5.20). In this statistic, we may clearly detect the fallacy of the assumption that still unused lands are readily available for farming. Because of the still rising crop yield rates per acre, the drop in total acreage has not yet meant a drop in total American agricultural productivity. But, considering the population growth and rising world need for food, we would certainly expect any really fertile farmland still available to be planted already, if such fertile acreage actually existed.

Most of the acreage being lost to agriculture each year consists of old, smaller farms of low productivity, abandoned because of their hilly terrain or poor, overused soil (Fig. 5.20). Hilly regions of thin, rocky soil do not lend themselves to modern or intensive agriculture, and farmers in such regions found it difficult to compete in the marketplace with the produce from farms of large fertile acreage. Yet, many old farmlands may yet have to be put back into production one day, if not already paved over or built on (Fig. 5.21) as the population continues to grow and food prices exceed even the high cost of crops grown on small farms of rough terrain.

Figure 5.20 (Top) Many mountainous farmlands like these in West Virginia have been abandoned because of difficulty of cultivation by mechanized methods on steep slopes and the susceptibility of such slopes to erosion as shown in the photograph at bottom. (U.S. Forest Service.)

Figure 5.21 Millions of acres of American farmland have been used for housing developments during recent years, thereby removing the land from use for food production. (Photo by author.)

It has been suggested that barren, rocky regions of soil too thin to plow may yet be made to yield crops by manufacturing and laying down a blanket of crushed minerals or other fine debris, and enriching it with organic matter such as compost or sludge left over after the treatment of urban sewage (see Chap. 2). Such artificial soils may yet be produced in quantity, indeed, may *have* to be but, for the present, their use would be even more expensive than current attempts to rejuvenate existing poor soils by extensive addition of similar organic materials.

FOOD WITHOUT SOIL

We have all heard statements regarding the vast future potential of the seas for providing food to the growing population. Unfortunately, increasingly intensive studies of marine food sources in recent years have shown sea food to be very limited when compared to the needs of man. Hopes of a vast marine food supply are more myth than fact. At present, the oceans provide little more than $\frac{1}{100}$ part of man's food consumption. In Chapter 4, we considered the reasons why there is little hope of increasing this fraction in the future.

Another possible food source consists of plants grown in fresh water, independent of soils. This process, called *hydroponics,* has held much hope among world food planners as a future solution to food needs when we reach the final limits of the soil's capacity for crops.

Hydroponics on any large scale is limited to the growth of algae. Quantities of larger, more complex plants are difficult to support without soil. The most useful alga used in hydroponics is the protein-rich genus *Chlorella.*

Hydroponic growth is a rather delicate procedure, requiring a more precise balance of environmental factors (light, temperature, and nutrients) than soil cultivation. As yet, large-scale hydroponics is very much in the experimental stage and entails so much effort that producing food products from algae is quite expensive. Nor have these products been found very palatable.

LIMITS

Adding new agricultural acreage by draining swamps, opening up roads to develop some still-unused lands, and extending irrigation to dry regions, all hold some future potential for increasing food supply. But the chief factors responsible for the enormous food production of the last three decades in the United States have been the use of chemical fertilizers, pesticides, and hybrid seed, plus ever-more efficient mechanical means of cultivation.

One of the most spectacular rates of crop increase per acre in the United States has been achieved in the growing of corn. The development of a variety of hybrid strains, each specifically suited to a particular latitude and climate have produced enormous yields and surpluses. This tailoring of corn to specific local requirements has been made possible by decades of genetic research and experimentation. Improved strains of wheat, sorghums, and rice have also been produced by American genetic and agricultural research. Some high-yield genetic varieties have been helping meet the food shortages in Asia, South America, and other crowded regions.

In the United States, crop yields per acre for corn and other plants, where hybrid strains have been very important, are still rising in some regions but leveling off in others. With some crops, there has been little or no increase in yield per acre for years. In general, we seem to be approaching the limits of the amounts of food that can be grown on American farm acreage, particularly in view of the limits of fertilizer addition imposed by the increasing danger of excessive quantities of nitrogen and phosphorous compounds passing into the waterways of the nation (see previous heading and Chap. 2).

Green Revolution

A great deal of attention has been given in recent years to the so-called green revolution, the name given to the use of new agricultural procedures, especially the use of chemical fertilizers and high-yield plant varieties, to farmlands in the underdeveloped countries. There is no doubt that use of high-yield rice and wheat strains is making major contributions to the food supply of these countries. However, the new grain varieties must be accompanied by high fertilizer use, and heavy pesticide application is also needed for maximum food production. These supplements require heavy economic outlays and present environmental dangers as considered earlier in this chapter. We should also bear in mind that the "green revolution" is limited to countries that have a good water supply.

In addition, the selective breeding of crop strains on which the green revolution is based, produces plant species with less generic diversity than the parent species and lower potential to adapt to new conditions, including future pests or diseases. The hybrid "miracle grains" being used increasingly in recent years to raise food yields actually increase the vulnerability of crops. Dependence of agriculture on such high-yield strains may lead to future crop failures and increased possibility of famines.

The green revolution seems to have given a temporary assist in the race to keep world food supplies from becoming even more inadequate for mankind. To many environmentalists, this phenomenon is vital in order to "buy time" with which to stabilize population growth. But in many countries, a rise in food production is likely to encourage more large families and further increases in population with little change in living standards or real improvement in diet for tomorrow's population.

Farmland and Food Distribution

There is a strong correlation between numbers of persons and land available to grow food. This can clearly be seen if we compare population densities in terms of agricultural land and in terms of standards of living. In the $\frac{2}{3}$ of the world where food supply is now inadequate, the ratio of persons to farmland is over 50 persons per 100 farm acres. In the $\frac{1}{3}$ of the world where food supply is adequate, the ratio is less than 18 persons per 100 farm acres. And this situation worsens each year because the average population growth in the underfed regions is about 2.2 percent a year compared to an average of about 1.3 percent of the adequately nourished third of the world.

The amount of food available per capita has been decreasing for years on a worldwide basis. During the two decades following the end of World War II, the world's total food production increased during most years (some recent years have seen no increase). But the control of disease and increase in infant survival rates during the same time period has meant such a rapid rise in world population that there has been a general *decrease* in food available per individual (Fig. 5.22).

The food requirements of the rapidly increasing population of most of the "underdeveloped countries" are rising above their own production

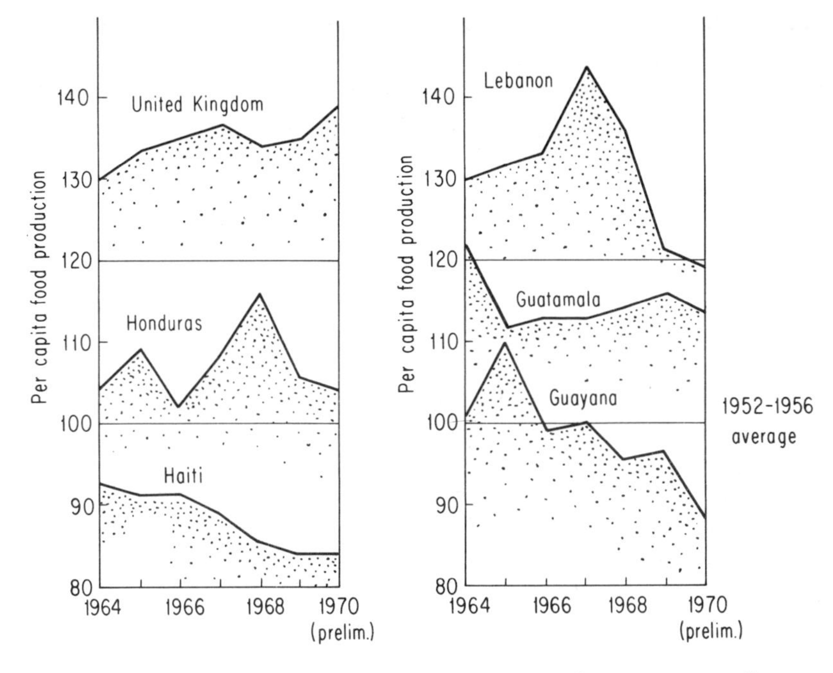

Figure 5.22 Trends in food production *per person* in five countries rapidly growing in population and one country, the United Kingdom, with a stable population.

capacities by many millions of tons of grain per year. At the same time the excess of grain produced by the United States, the only major donor of food to these countries, has been rising at a slower rate.

Numerous scientists believe that well before the year 2000, rising population and food needs will have far outstripped available food in many regions, and widespread famines will occur. The manner of distribution of the food surpluses that are available at that time may largely determine where starvation will occur.

There are increasing indications that we are approaching the overall limits of the planet as far as food supply for humanity is concerned. It is far from certain that the earth's fields and seas can provide enough food for the six or seven billion inhabitants projected for the year 2000 in spite of the "green revolution" and other agricultural achievements.

Every three years the human race is increased by an amount equal to the entire population of the United States. Every six weeks a number equal to the population of Sweden is added to humanity. It is clear that we have long since passed the limits of the planet to adequately feed all members of the human race. Over $\frac{2}{3}$ of the human inhabitants of the planet are malnourished or underfed.

The United States is in the position of the child born with a silver spoon. The youthfulness of its soil, relative sparseness of population, and its development during an age of growing mechanization have permitted the growth of a highly productive agriculture, essentially an agricultural industry capable of producing rich surpluses. It is difficult for us as part of the fortunate, well-fed quarter of the world to believe that most of the humans now being born will suffer hunger and an early death from starvation or malnutrition-related illnesses.

Suggested Readings

BORGSTROM, GEORG, *Too Many*, Collier-Macmillan, Toronto, 1969.

DALE, TOM, and V. G. CARTER, *Topsoil and Civilization*, University of Oklahoma Press, Norman, 1955.

DONAHUE, ROY L., *Soils: An Introduction to Soils and Plant Growth*, 2nd ed., Prentice-Hall, Englewood Cliffs, N.J., 1965.

MCNEIL, M., "Lateritic soils," *Scientific American* offprint No. 870, 1964.

PADDOCK, WILLIAM, and PAUL PADDOCK, *Famine 1975!*, Little, Brown, Boston, 1967.

ROBINSON, H. F., "Dimensions of the World Food Crisis" in *BioScience*, Vol. 19, No. 1, 1969.

Rotarian (magazine, special issue), "This Hungry World," June, 1969.

WHARTON, CLIFTON R., "The green revolution: cornucopia or Pandora's Box?" *Foreign Affairs*, Vol. 47, 1969.

6
Limits

Photo courtesy of USDA Soil Conservation Service.

THE MYTH OF INEXHAUSTIBILITY

When Europeans first lived in what is now the United States, it was the home of some million Indians who had lived here for thousands of years, causing very little visible effect on the land surface. They probably altered the American landscape less than did the beaver. Now there are over 200 million persons in the same space, and the land surface has been altered virtually everywhere.

Centuries ago the entire eastern half of the continent, and a large part of the west, was covered with forests. Regions now virtually treeless, like most of Ohio and Indiana, were then the sites of magnificent forests with trees reaching 100 feet in height. These original forests were among our first land resources to be depleted as millions of acres of trees were cut down and burned merely to clear the land for early farming. Later, most of the forests that remained were cut over to produce lumber for growing towns (Fig. 6.1). The virgin forests of our country were destroyed with little concern for the future. To early settlers, the land seemed virtually inexhaustible, so vast was it, when compared with the European home countries.

For nearly a century after American independence, vast untouched tracts of land still lay beyond the frontier, and the idea of unlimited resources became well-ingrained in the American attitude. This historic myth of inexhaustibility of the land can still be detected in the wide-

Figure 6.1 Destruction of forests, like these in Oregon, has been a typical by-product of the United States' assumption of virtual inexhaustibility of its resources. (Photo by U.S. Forest Service.)

spread unwillingness of Americans to accept restrictions on the use of resources, whether trees, fossil fuels, or metallic ore deposits, or to face the problems of air and water pollution. Let us consider next the problems of ore resources.

ORE DEPOSITS

Ore deposits are masses of minerals from which metals can be obtained commercially. Every ore body is a rock in which there is a concentration of a metallic element *greater* than that of the average in the earth's crust. Most of the various industrially-used metals, such as copper, zinc, tin, and lead, are found in the earth's crust in very small amounts, each comprising less than $\frac{1}{100}$ of one percent. An ore of one of these elements is a rock *greatly enriched* in the metal. Lead, for example, may be enriched a thousand times over its average crustal percentage to reach ore grade.

Of the useful metals, only iron, aluminum, and magnesium are found in substantial amounts in the crust as a whole. But even ores of those elements have been enriched. A low grade ore of aluminum, for example, with about 25 percent metal, has been enriched to several times over the concentration of aluminum in average crust rock.

We can consider all of the metals that are less abundant in the crust than magnesium to be scarce metals. It is surprising to realize that such metals as copper, lead, nickel, and zinc, all of which have large uses, are rare in terms of total abundance in crustal rocks. Other "scarce" metals include tungsten, titanium, manganese, chromium, molybdenum, and

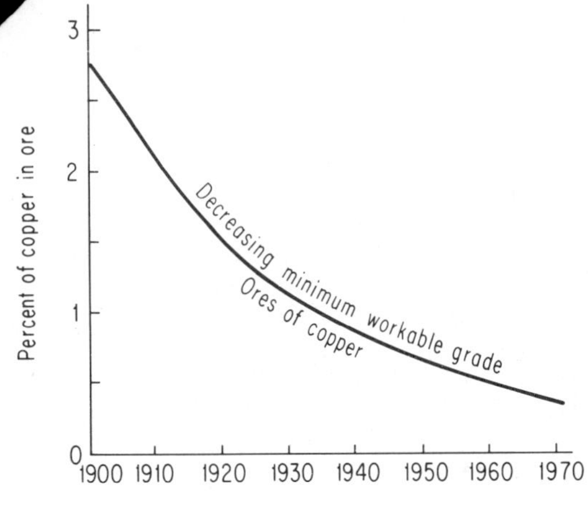

Figure 6.2 Rich ores of many metals are being exhausted, and lower grades of these ores must be used as metal sources. Copper is a prime example of ever-decreasing percentage of metal in ores being mined.

mercury. Many of these are required only in small amounts for steel manufacture or for other industrial purposes, yet may be critically important. Some of these have abundances adequate for the distant future, but others already show serious depletion, with much lower grade ores now being used than were used a decade ago (Fig. 6.2).

The Time Element

Regardless of their mode of origin, all ore deposits are rocks in which metallic elements have been concentrated by geologic processes, operating over a very, very long time by human standards. The time required to produce a particular ore deposit may have been centuries or thousands

Figure 6.3 Many mining towns in the western United States, like this one in Nevada, were abandoned when ores were depleted. (Photo by author.)

of years, and some deposits probably required hundreds of thousands or millions of years to build up.

The earth contains fixed amounts of the metal-bearing minerals so that these ore deposits are nonrenewable resources like the fossil fuels considered in Chapter 3. All ore bodies have their limits and will be mined out eventually if worked long enough. With a rate of formation so slow, there can be no "second crop" of minerals to harvest. When an ore deposit is mined out, all that remains is a hole in the ground. Many great ore deposits have reached this condition with only ghost towns left remaining where there had once been active mining communities (Fig. 6.3).

In many cases, lower grades ores have had to be used in place of depleted high grade ores. Consider the case of the Mesabi iron ores of Minnesota, the chief source of iron in this country for nearly a century. They consist of hematite (iron oxide) that was formed by a very slow alteration process from the original, ancient sedimentary rock, a mixture of silica and iron called *taconite.* The alteration was caused by downward-moving waters that caused the oxidation of the iron of the taconite and the leaching out of the silica and other substances. During a period of thousands of years, water percolated downward hundreds of feet, altering the taconite and leaving rich concentrations of hematite that were readily mined by power shovels from open pits (Fig. 6.4).

The great demand for iron during World War II greatly depleted the Minnesota hematite ores. As open pits grew deeper, the bottom of the alteration zone was reached in parts of the Mesabi region. The near

Figure 6.4 Increasing demand for iron ore results in numerous open pit mines, like this giant pit in Minnesota. (Photo by author.)

exhaustion of the rich ores seemed to present a serious problem of iron supply until new technological means were found to use the parent taconite layers themselves as iron ores. Taconite is now widely used in blast furnaces after being put through an artificial concentration process called *beneficiation.*

Scarcity of Ores

Growing population and expanding per capita use of metals have meant a rapid rise in the tonnage of minerals required each year by mankind. The annual average use of mineral substances (metallic and nonmetallic) in the United States now exceeds 10 tons per person.

As shafts deepen and pits widen, obtaining these minerals becomes more difficult. As the largest and richest ore deposits near exhaustion, it becomes necessary to use ever-leaner deposits.

Of the three most common metals in the crust, aluminum is most abundant. Aluminum ores have been mined for so short a time that rich deposits are still readily available, although they are chiefly outside the United States in tropical regions of Africa, South America, and elsewhere. There is little chance of running out of aluminum-bearing minerals, but once the bauxites are depleted, clays may have to be used, the cost of extracting aluminum from ore, already much higher than the cost of iron extraction, will undoubtedly rise rapidly.

Iron is the second most abundant metal in the crust. So much iron has been mined in this country that the richer deposits have been greatly depleted, and we must depend on leaner ores, like the Minnesota taconite, supplemented by foreign ores. Nevertheless, there is probably little possibility of "running out" of iron for centuries to come because of its overall abundance in the crust and the rich deposits that still remain in Canada, South America, and elsewhere.

Magnesium will certainly present no problems of future supply because of its overall abundance in crustal rock and in sea water, and its relatively low rate of use.

It is the remaining, scarcer metals that present serious problems of depletion. Copper, lead, nickel, and uranium (see Chap. 3), for example, each comprise less than 0.01 percent of the earth's crust; titanium, manganese, and chromium are only slightly more abundant.

The concentration of some metals varies more or less continuously from very high grade ores down to mere traces. This is the case for aluminum and iron. Some copper ores also exhibit complete gradation of concentration percentages. But most other scarce metals do not show such continuous gradation from rich ore to average crustal rock. Most ore deposits of lead, zinc, nickel, manganese, tungsten, and molybdenum have only narrow zones of transition from the enriched ore body to the

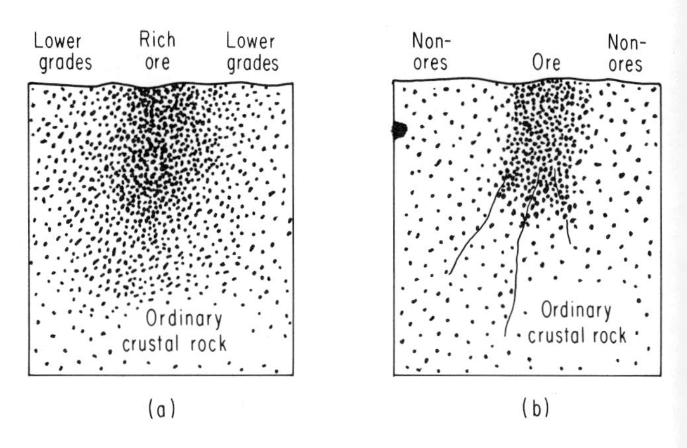

Figure 6.5 (a) Some ore deposits are transitional into surrounding ordinary crustal rock so that ever-lower grades of ore can be mined if demand requires. (b) Some metals occur in ore deposits that end sharply with little transitional, lean-ore zones. These are the metals the supplies of which are most likely to be exhausted.

original, unenriched adjacent crustal rock which may have only a tenth the percentage of metal of the ore body (Fig. 6.5). These metals show sharp drop-offs in metal percentages from the relatively rich ore to rocks with metal content too small to be used except in ultimate desperation. Thus, for many of the scarcer metals, we face the real possibility of running out of usable ore deposits.

Metals in the Oceans

The oceans are a giant repository for elements, including metals, which have been brought from the land by rivers and streams and left dissolved or suspended in seawater or precipitated or deposited on the sea floor. Thus, the metal content of the oceans is due mainly to the erosion of the land. Because it has been found to contain almost all of the metals and other elements used by man, seawater can be considered a potential source of the metals needed in the future. Unfortunately, the concentration of all but a few of these elements is so tiny that it is now economically unfeasible to extract them from seawater (see Chap. 4).

Future technology may change the picture a bit but, at present, we must conclude that metals in seawater are at much too low a concentration to be recoverable in practical terms or at a profit. One notable exception is magnesium. It is one of the most concentrated elements found in seawater and can be readily recovered by precipitation brought about by chemical action.

New Finds

What are our chances of finding another huge concentration of ores in this country like those of Butte, Montana? Very small. This country has been probed and explored by more geologists than any other on earth. New outcrops of ore deposits may yet be discovered by exploration, but most will be in the remote regions of the world.

A potential source of ore for the future are the "blind" ore bodies, deposits rich enough to mine but that do not reach the surface. Unexposed as they are, these bodies usually must be sought by complex geological methods and drilling programs. The deeper the body, the more likely an expensive exploration program will be required for discovery and delimitation.

In addition to the expense of exploration for presently unknown or hidden ore bodies, there is the problem of the time required to prepare for commercial use of newly discovered deposits. This "lead time" may involve many years during which excavations are begun, equipment obtained and installed, and roads built. Five years is about the average time between discovery of a major ore deposit and commercial use of the metal.

We should bear in mind that remote regions have been growing scarce while geologists are more numerous than ever. Yet, ores of most metals are being used up much more rapidly than new deposits are being discovered.

Technological Advances

There is no question that advances in science and engineering have enabled us to use the Minnesota taconite, and the very low-grade copper ores of the Southwest, and generally to offset the economic costs of having to mine deposits of such low-grade. The question is can this continue? At the present time, there is considerable doubt that technology will be able to keep down the costs of mining copper ores and other ores that grow leaner year by year. There is also a new awareness of the previously hidden and growing evironmental costs of the technology of mining and processing low-grade ores (see below).

It is assumed by some optimistic planners that as demand for metal increases, mining companies will be able to move on to lower and lower grades of presently noncommercial mineral deposits. These lower-grade deposits will become ores as demand requires and technological advances permit, especially if nuclear plants provide cheap, abundant energy.

Even if nuclear energy was safe, available and very cheap, it is not clear how its use would permit the mining of all of the world's low-grade ores. The breakup of crustal rock by nuclear explosions has been suggested but seems impractical, considering the complexity of most ore bodies and the radiation dangers. It may be more realistic to think in terms of using nuclear energy indirectly, in the form of low cost electricity, to run especially powerful mining machinery and giant processing plants. But cheap nuclear energy is no longer as promising as it once seemed. In Chapter 3, we considered the limited availability of uranium fuel for atomic fission and the lack of progress in atomic fusion; in Chapters 1, 2, and 3 we considered the dangers of radiation and thermal pollution from nuclear power plants. For these reasons, there are grave questions of how important nuclear power will be in future decades in using low-grade ores.

Clearly, there *are* limits to the achievements of science and technology. To expect our scientists and engineers to wring endless supplies of metals and other substances out of the crust in order to meet soaring expectations of ever-growing, more affluent populations is to live in a world of unrealities.

Environmental Damage from Metal Consumption

The effect of mining on a landscape is invariably unsightly. Any traveler in the mountains of Colorado, Idaho, California, or other western states has seen the old mine shafts, decaying buildings, great piles of tailings (wastes), and the streams stained yellow and brown from mine leakage long after the mines have closed (Fig. 6.6). Vast, barren piles of gravel and other evidences of hydraulic gold mining can still be seen along western rivers as much as half a century after mining ended (Fig. 6.7).

The rapid increase in size of unsightly open pits (Fig. 6.8) and other kinds of metal mines in this country is caused by two factors: (1) the rising yearly demand for almost all metals and (2) the exhaustion of rich ores, which requires extractions of larger amounts of lower grade ores. To illustrate the latter trend, consider that a copper ore with a metal content of one percent is exhausted, and ore with a content of half of one percent must be used; twice as much ore will have to be removed. Or, if hematite ore of 60 percent iron content is exhausted and taconite of 30 percent iron is substituted, again twice as much ore must be removed to obtain the same amount of metal, and the open pits will increase in size at double the former rates (Fig. 6.9).

Open pit mines for iron and copper have become major landscape features in several states (Fig. 6.10). The Bingham Canyon pit in Utah

Figure 6.6 Abandoned mine tailings and debris mark the sites of former mines in Colorado. (Photo by author.)

Figure 6.7 Gravel banks mark the location of placer gold deposits in Colorado; mining was discontinued decades ago, but landscape damage remains. (U.S. Forest Service.)

Figure 6.8 Abandoned iron mine workings. This open pit is located in Minnesota. (U.S. Forest Service.)

Figure 6.9 New mines deepen. Open pit copper mining in Arizona. (Photo courtesy of the Anaconda Company.)

Figure 6.10 Open pit copper mine complex in Arizona. (Photo courtesy of the Anaconda Company.)

has become one of the largest man-made depressions in the world from which nearly a billion tons of rock may be removed in a single week.

In Chapter 3, we considered the landscape damage of strip mining for coal that is similar to open pit mining for ores and the dangers of radioactive pollution recently recognized in the tailings of the western uranium mines.

Many metal processing and refining plants damage the surrounding landscape by stack emissions and discharges into streams and lakes. Copper smelters in particular have had a major effect on local landscapes because of damage to vegetation caused by sulfurous pollutants discharged into them during the processing of copper ores (Fig. 6.11).

Figure 6.11 (Top) Copper smelter in Arizona. Sulfurous air pollutants damage vegetation and other life for miles around. (Bottom) Copper smelter caused extensive vegetation destruction and soil erosion in this region of Tennessee. (U.S. Forest Service.)

Ores and Future Energy

There can be no doubt that eventually metals will have to be extracted from mineral deposits that carry only fractions of the amounts now found in commercial ores. The larger the amount of rock mined and processed for each ton of a metal, the greater the amount of energy required per ton. Producing the great amounts of additional energy that will be needed to extract and process these vast future tonnages are likely to have severe environmental consequences.

If the energy is to come from burning fossil fuels, then we must be prepared for strip mining (Fig. 3.10) at ever-accelerating rates and for the increased air pollution and thermal pollution from the power plants. These environmental consequences of power production were considered in Chapter 3 and in other chapters. If the energy is to come from nuclear sources, then we must assume the development of breeder reactors and must be prepared to undertake strip mining of great new tracts of landscape in order to obtain low-grade uranium-bearing deposits. There will be greater amounts of thermal pollution from the nuclear plants than from fossil fuel plants, as well as a great rise in the level of potentially dangerous radiation in our environment.

The environmental costs of metal used and of processing very lean ores cannot be ignored, especially when optimistic social planners envisage that technology will permit extracting metals from common rocks like granite, or from the sea itself, through the availability of cheap, unlimited nuclear energy. Optimism of this sort is the result of ignorance concerning the environmental facts of life or of naive or mystical assumpions.

Only large-scale solar energy use or, perhaps, atomic *fusion,* hold any real promise of cheap and safe (or relatively safe) energy in vast quantities for the future. Neither has yet been developed and may never be. All existing energy sources, including nuclear fission, have serious limitations. For a discussion of these limitations and environmental consequences of energy use, review Chapter 3.

American Versus World Ore Use

The United States became a world power largely because of a relatively small population in a large fertile land underlain by vast resources of metals and fossil fuels. These resources included nearly 30 percent of the world's iron ores, nearly 25 percent of its copper ores, and nearly 40 percent of its coal. But now the country stands first among nations in the rate at which it is using up its own mineral resources. In its ever-growing need for industrial raw materials, the United States has consumed more minerals in the past 35 years than *all* of the people of the world put together, had used previously, *from the beginnings of history.*

In spite of its own mineral wealth, the United States depends on foreign supplies for a good part of its needs. The reserves of ores and other valuable mineral concentrations in this country are quite well-known by now and measured (at least approximately). Discoveries of new ore deposits are still going on but at a generally declining rate. As every geologist knows, we are long past the peak of discovery, and no new major mining districts have been found in this country in the last half century of mining exploration.

Americans are the greatest consumers the world has ever known. The American standard of living has become a goal of many of the peoples of the world. Underdeveloped countries hope to achieve this goal for their citizens by rapid industrialization. But this is often an unrealistic goal when considered in terms of the resources that would be required.

Let us consider what would be needed in the way of metals in order to fully industrialize and electrify all of the nations of the world and bring their peoples up to the present American standard of living by the year 2000. For example, about 200 times as much copper would have to be extracted from the ground as is now mined annually, 400 times as much lead, and 500 times as much tin as is now produced annually would also be needed, and similar huge quantities of other metals would be called for.

Of course, there is no realistic likelihood of industrializing all of the world or raising humanity's living standards to anywhere near American levels in two or three decades, especially since American consumption will also continue to soar. Nor is it likely that our environment could withstand world industrialization because of the overwhelming pollution it would cause, as considered in previous chapters, and because of the production of solid wastes, to be considered next.

SOLID WASTES

The population of the United States has been growing in its affluence and per-capita consumption faster than in numbers. Just as we are polluting our rivers and atmosphere with the liquid and gaseous wastes of affluent living, the production of solid wastes has now reached truly incredible proportions and comprises one of our greatest environmental problems.

The average American now produces nearly a ton of solid waste per year. These wastes include a great variety of substances, some biodegradable, others nonbiodegradable. Some are combustible, some not, and all burning causes air pollution (see Chap. 1). Many wastes from industry and hospitals are poisonous. Others yield carcinogenic by-products. Much solid waste produces foul odors; little waste is totally inert; almost

Figure 6.12 Waste disposal in the United States too often takes the form of open dumps that pollute rivers and damage landscapes like this river edge disposal in Kansas. (USDA Soil Conservation Service.)

all of it is unpleasant to the eye. Certainly all solid waste occupies space and, for many cities, towns, and countries, the problems of where and how to dispose of the ever-increasing bulk has become critical.

The strong American tradition of individual freedom is too often taken as license to dump waste wherever convenient. For centuries, we have thrown trash over cliffs, beyond fences, off roadsides, into woods, rivers, and lakes, in short, anywhere out of sight of the discarder (Fig. 6.12). This sort of dumping, as well as burning waste, is becoming unacceptable in many regions because of the resultant pollution or disfigurement of the landscape and the decline of available space isolated from habitation. There is simply not enough land any longer for us to dump all of our wastes out of sight (and smell) of others.

Dumping

Many American cities located near the seashore, find the ocean the most convenient place to dispose of their solid wastes. New York, for example, has dumped some 15 million tons of wastes into the ocean during a 10-year period. For most inland American towns and cities, ocean dumping is of little practical use. Even for coastal communities, ocean dumping may be more expensive than using land dumps or landfill operations because of the use of loading facilities, barges, and tugboats. But the practice is growing yearly, adding to the rapidly increasing pollution of the seas, a problem discussed in Chapter 4.

Open dumps are the most primitive land repositories. Unfortunately, rural residents often practice dumping on an individual basis and in

Figure 6.13 Open dumps breed vermin and damage landscapes. (Photo by author.)

haphazard locations. Most small municipalities in this country use primitive open dumps and often burn their dumps periodically in order to decrease the quantity and hasten the deterioration of the wastes. But there is increasing opposition to burning, and many counties and a few states have outlawed the practice because of the resultant air pollution.

Even without open burning, open dumps present major environmental hazards (Fig. 6.13). They are breeding grounds for vermin and disease-carrying insects. Water percolating through dumps dissolves or leaches out a variety of undesirable or toxic substances, and the *leachate* often contaminates nearby rivers, and lakes, and groundwaters below the surface (Fig. 6.14 and 6.15). It often contains toxic substances such as lead, mercury, arsenic, and potentially dangerous viruses such as the one which causes infectious hepatitis. In spite of these hazards, tens of thousands

Figure 6.14 This open dump in Tennessee is polluting both surface and ground water. (USDA Soil Conservation Service.)

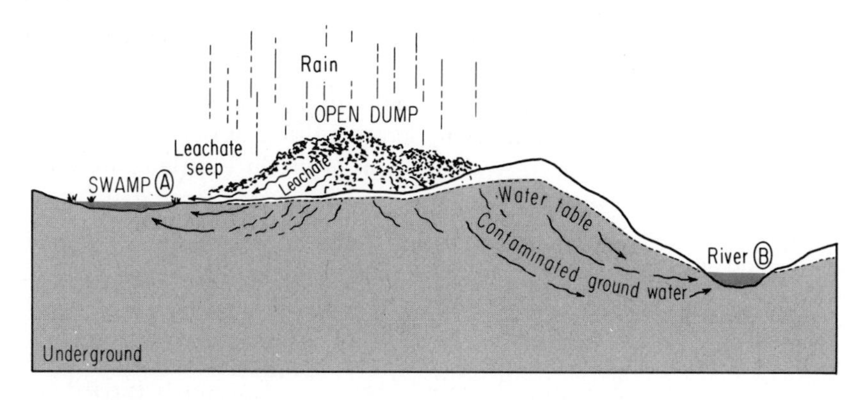

Figure 6.15 Open dumps contaminate nearby surface water in a swamp (A) by leachate migration and (B) contaminate ground water that may further contaminate other surface waters.

of open dumps are being operated in the United States. California alone had over 500 in 1971.

A more advantageous method of dumping waste is the *sanitary landfill* in which the waste dumped during each day of use is covered by a layer of soil, clay, or sand. Usually the waste accumulation is compacted somewhat before being covered. It is the clean soil or other added material that makes the landfill sanitary. This cover is also compacted and, in theory, the layers act as barriers to isolate the gases and leachates produced by the deterioriating waste from the atmosphere and from adjacent water supplies (Fig. 6.16). Ideally, such landfills can be used for parks, golf courses, playgrounds, parking lots, or other useful purposes.

In practice the distinction between a sanitary landfill and an open dump is not always clearcut. The intervening clean earth barriers are

Figure 6.16 Sanitary fill operation. Bulldozer is covering recently dumped waste, with clean earth material. Older layers of waste and earth shown in cutaway view.

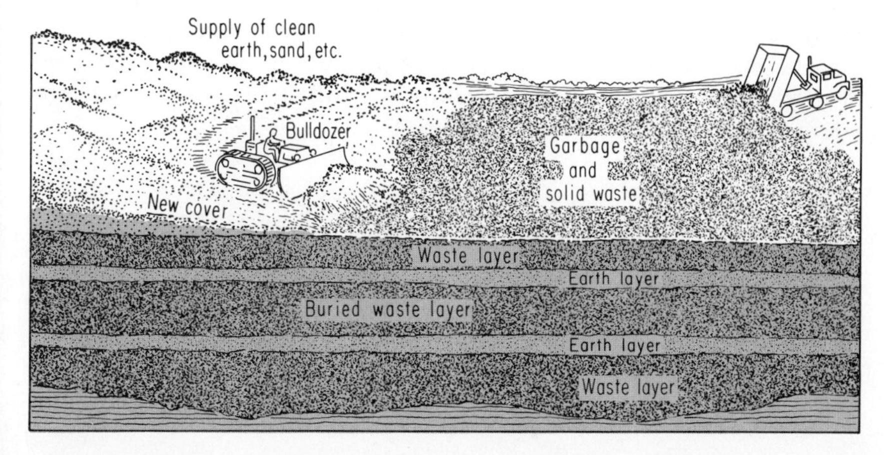

frequently inadequate to prevent the leachate flow and resultant contamination of adjacent water bodies. Decomposition of landfills commonly produce gases which may be smelly or even explosive, as in the case of methane caused by anaerobic decomposition of organic matter.

Even a landfill constructed so carefully as to produce little or no water pollution, alters the nature of the region in which it is located. In many areas, for example, sanitary landfills and other dumps are displacing marsh lands essential for the support of bird and aquatic life.

Incineration and Pyrolysis

Incineration is a procedure in which solid wastes are reduced to inert residues by high temperature burning. True incineration is carried out under carefully controlled circumstances in appropriate installations. The residue from an incinerator may be reduced to 10 percent or less of the volume of the original waste and can be disposed of in landfills much more safely because of the destruction of much of the undesirable substances.

The greatest problem associated with incineration is the pollution of the atmosphere it causes. In the United States alone, increasing concern for the quality of the air has led to the closing of thousands of municipal incinerators in the past few years, and of countless apartment house incinerators, and other small installations.

There has been much discussion by American environmentalists of the possibilities of using the heat of incineration to generate steam and produce electrical power. This recovery of energy has been practiced in some parts of Europe for many years. For example, Paris has been running steam-electric plants by burning solid wastes for the last 50 years. Over a million tons of city refuse is converted into electricity each year, and the left-over steam heats hundreds of large buildings in the city.

Pyrolysis (destructive distillation) is a little-used method of waste disposal that has enormous potential for the future. Pyrolysis is a process of decomposing waste materials at high temperatures but in the absence of air. There are no pollutants discharged into the air, and a number of useful by-products can be recovered, including methane gas, oils, tar, carbon, and various chemicals. The apparatus for pyrolysis has been developed, but in 1971 the procedure was still being carried out only on an experimental or small-scale basis.

THE NEED TO RECYCLE

No one ever really consumes anything. Whether it is an automobile, a radio, or a can, the object is used for a time and then usually dumped somewhere. *Recycling* is an increasingly important alternative to dumping which can serve to reduce the amount of waste accumulation, and also allow the reuse of materials, and decrease the drain on our shrinking resources.

The degree to which metals are presently reused or recycled varies widely with the metal and the location of the metal article when its usefulness is over. A relatively large amount of lead is recovered and reused because most dealers in automobile batteries reduce the cost of a new battery when an old one is returned. The dealer returns the exchanged battery to the manufacturer who recovers the lead for use in batteries.

The abandoned automobile is a particularly obvious example of needless waste of materials that could have been recycled. Until a few years ago, the value of the scrap steel and other parts of an old car was great enough to encourage its owner to sell it to a junkyard for scrap. Because of labor costs and other factors it now frequently costs more to move a car to a scrap steel dealer than it is worth. As a result, most automobiles are abandoned, over half a million each year, each representing several tons of iron ore and of coal that had to be used to produce the car, resources that are gone forever (Fig. 6.17).

Figure 6.17 Most automobiles in the United States are eventually abandoned and left to rust, thereby losing iron that could be recycled to help decrease iron depletion rates. (USDA Soil Conservation Service.)

For another prime example of wasted resources, we can turn to the vast numbers of containers for liquids manufactured in the United States. Most soft drinks and beers used to be sold in returnable deposit bottles. Two decades ago, the average deposit bottle made over 25 round trips from manufacturer to consumer and back. Today deposit bottles have largely been replaced by nonreturnable, disposable bottles and cans. Both represent a drain on natural resources, with the convenience of cans (usually aluminum) especially dearly bought, considering the drain on aluminum ores (Fig. 6.18). Use of aluminum disposable cans also contributes to rising pollution levels because of the large amounts of electrical energy needed for aluminum manufacture (see Chap. 3). Recent campaigns to collect and recycle aluminum cans are helpful because the energy required to produce a ton of aluminum from scrap is considerably less than that required to produce a ton of aluminum from bauxite ore. But in 1972 the number of aluminum cans actually recycled represented only a very tiny fraction of the cans being produced.

Figure 6.18 Most aluminum cans in the United States are discarded, often becoming a blight on the landscape. Only a very small percentage were being recycled in 1972, a procedure that could greatly decrease the amount of electrical energy used to refine aluminum ores. (USDA Soil Conservation Service.)

In fact, the number collected was considerably less than the increase in the number of cans over those produced the previous year.

It is clear that major changes in policy will have to be made in order to increase the amount of used metals recovered and recycled. Recycling of metals is an inevitable next stage in man's long use of metals. The sooner we arrive at this stage the better for the conservation of our remaining ore deposits and for our environment.

It is long past the time that waste be considered as valuable material, even as a displaced resource. Automobiles, metal containers, and other commonly discarded objects should be treated as substances that *must* be recycled. The sale of a car should include a provision for its eventual return to be recycled. Container manufacture needs to be redirected to reduce discards and ensure a high percentage of reuse. This can mean a return to deposit bottles, possibly even the use of deposit cans, on which the deposit fee is large enough to ensure the return of most of the containers sold. The goal should be the recovery of the basic materials, especially the metals, used in most products.

IS GROWTH GOOD?

The birth rate in the United States has been dropping in recent years (about 17 per 1000 in 1970) but is still higher than the death rate (about 13 per 1000) so that our population continues to rise. With an average of about 60 persons per square mile of land surface we are still generously endowed with land compared to most of the world (India's density is about 440 persons per square mile). This fact has been one basis for our long-held feeling that growth is good, even vital to our national welfare. New York State and California have vied with each other for years to see which could achieve the largest population. In New York City, utility company construction signs read: "Dig we must for a growing New York," and most passersby accept the need, in spite of the crowding and rapid deterioration of the city evident to almost anyone.

In the United States we have long espoused the philosophy that growth is good. But in recent years, with the frontier long gone, it has become increasingly apparent that we are losing more by growth than we can possibly gain. Problems of water supply are already widespread and threatening drastic shortages. Air pollution affects almost everyone's health and longevity. DDT and other poisons intended for insects have become widespread and are building up in our bodies.

During the past quarter century, the population of the nation has increased by less than 50 percent, but total energy use has increased by more than 200 percent (see Chap. 3). Increases in production of many specific pollutants during the same interval have been even more spectacular. For example, nitrogen oxide emissions by automobiles in this country increased about 600 percent, and phosphate levels in sewage also increased some 600 percent, all in 25 years.

The final Report of the Commission on Population Growth and the American Future, released in 1972, made it clear that "population growth is one of the major factors affecting the demand for resources and deterioration of the environment." According to the Commission, "no substantial benefits would result from continued growth of the nation's population."

United States and World Problems

It is quite clear now that population growth in the United States has had a disproportionate effect on environmental deterioration and on depletion of the world's resources. In relative terms, it has been estimated that each American utilizes about 50 times as much of the nonrenewable resources of the world as the average citizen of underdeveloped countries like India.

Already many scientists, including the author of this book, doubt that the entire population of the world, even if it grew no larger, could *ever* live at the current level of United States affluence considering the resources available and the inevitable pollution consequences. Many environmentalists believe that no more than a quarter of the present world's population could ever be supported at American levels of consumption of food, resources and energy, and that all efforts should be made to *decrease* world population even as we try to increase the quality of life in the poor nations. It has become a choice between more people at a lower level of existence or fewer people at a higher level, with more opportunities for comfort and fulfillment.

The Major Causes of Environmental Problems

Most of the environmental problems we have treated in this book can be attributed to one or more of the following trends:

1. Overpopulation of the planet
2. Excessive per-capita consumption in the highly-developed countries
3. Detrimental technological "advances"

In many regions, including most of the underdeveloped countries, the overpopulation factor and its resultant drain on food supply and availability of jobs and capital is the most critical cause of low standards of living and environmental problems. In other regions, especially in the

United States and many European countries, overpopulation aggravates the loss of environmental quality, but it is the rapid rise in per-capita consumption of resources, coupled with the growing trend toward waste-producing and nonbiodegradable disposables, as the result of mass-production technologies, that have been most responsible for environmental damage.

Who is actually to blame for environmental deterioration in the United States? Is it individual carelessness, or is it corporate and government policy that steers us ever closer to a state of irreversible environmental damage? In the opinion of the author, it is collective entities, both corporate and governmental, rather than the average citizen, that bear the primary responsibility for our failing state of environmental health. Citizen's efforts toward resource conservation are too often futile in the face of corporate or governmental policies to the contrary.

Suggested Readings

CLOUD, PRESTON, "Our disappearing earth resources," *Science Year,* 1969.

COMMONER, BARRY, *The Closing Circle,* Knopf, New York, 1971.

GRINSTEAD, ROBERT, "The new resource," *Environment,* Vol. 12, No. 10, 1970.

HULETT, H. R., "Optimum world population," *BioScience,* Vol. 20, No. 3, 1970.

JARETT, H. (ed.), *Environmental Quality in a Growing Economy,* Johns Hopkins Press, Baltimore, 1966.

MEADOWS, D. D., J. RANDERS MEADOWS, and W. BEHRENS, *The Limits to Growth,* Universe Books, New York, 1972.

National Academy of Sciences, Committee on Resources and Man, *Resources and Man,* W. H. Freeman, San Francisco, 1969.

OSBORN, FAIRFIELD, *The Limits of the Earth,* Little, Brown, Boston, 1953.

PARK, C., *Affluence in Jeopardy,* Freeman, Cooper, San Francisco, 1968.

STEWART, GEORGE R., *Not So Rich As You Think,* Houghton Mifflin, Boston, 1968.

WHITE, LYNN, "The historical roots of our sociological crisis," *Science,* Vol. 155, 1967.

Appendix

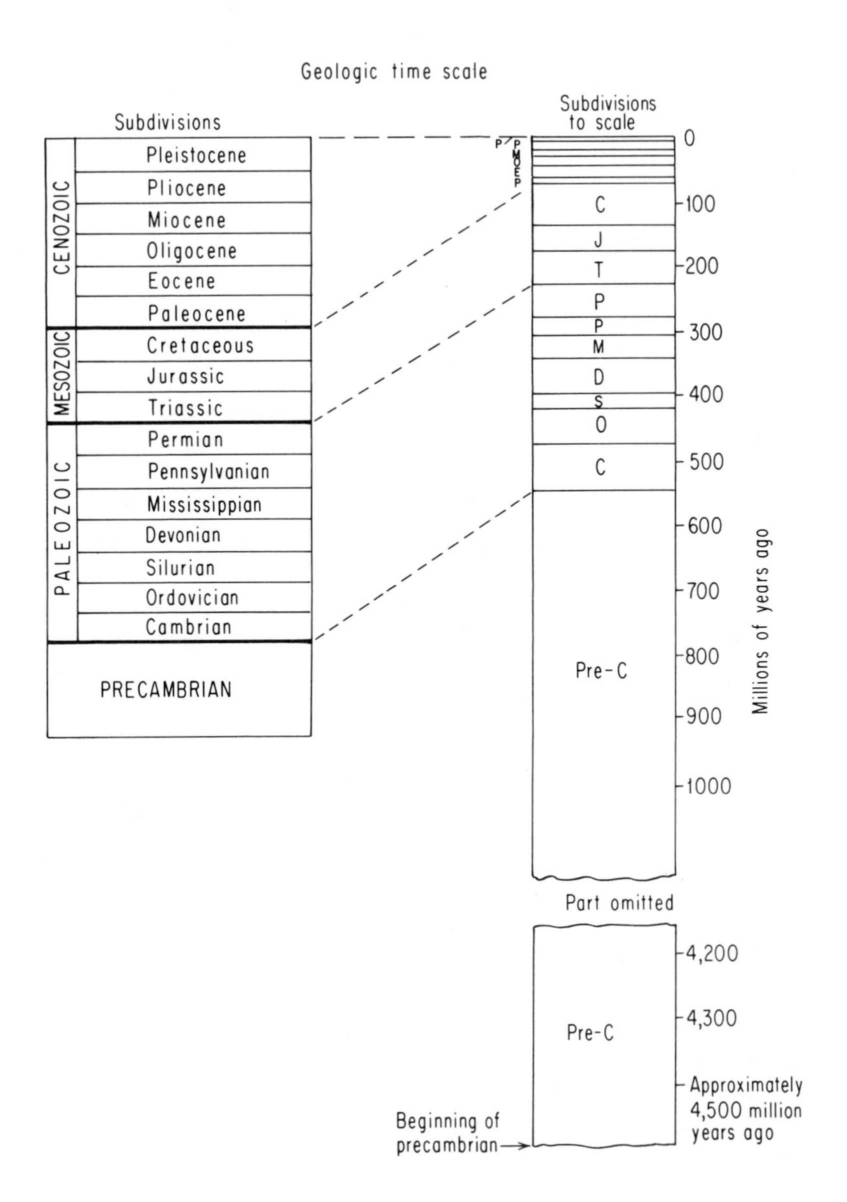

Geologic time scale
Subdivisions
CENOZOIC
Pleistocene
Pliocene
Miocene
Oligocene
Eocene
Paleocene
MESOZOIC
Cretaceous
Jurassic
Triassic
PALEOZOIC
Permian
Pennsylvanian
Mississippian
Devonian
Silurian
Ordovician
Cambrian
PRECAMBRIAN
Subdivisions to scale
P
P
M
O
E
P
C
J
T
P
P
M
D
S
O
C
Pre-C
0
100
200
300
400
500
600
700
800
900
1000
Millions of years ago
Part omitted
Pre-C
4,200
4,300
Approximately 4,500 million years ago
Beginning of precambrian

Region or Country	Population Estimates Mid-1971 (millions) †	Annual Births per 1,000 Population ‡	Annual Deaths per 1,000 Population ‡	Annual Rate of Population Growth (percent) °	Number of Years to Double Population □	Annual Infant Mortality (Deaths under one year per 1,000 Live Births) ‡	Population under 15 Years (percent) ▶	Population Projections to 1985 (millions) †
WORLD	**3,706[1]**	**34**	**14**	**2.0**	**35**	**—**	**37**	**4,933**
AFRICA	**354[2]**	**47**	**20**	**2.7**	**26**	**—**	**44**	**530**
NORTHERN AFRICA	**89**	**47**	**16**	**3.1**	**23**	**—**	**45**	**140**
Algeria	14.5	50	17	3.3	21	86	47	23.9
Libya	1.9	46	16	3.1	23	—	44	3.1
Morocco	16.3	50	15	3.3	21	149	46	26.2
Sudan	16.3	49	18	3.2	22	—	47	26.0
Tunisia	5.3	45	14	3.1	23	74	44	8.3
UAR	34.9	44	15	2.8	25	118	43	52.3
WESTERN AFRICA	**104**	**49**	**23**	**2.6**	**27**	**—**	**44**	**155**
Dahomey	2.8	51	26	2.6	27	110	46	4.1
Gambia	0.4	42	23	1.9	37	—	38	0.5
Ghana	9.3[4]	48	18	3.0	24	156	45	14.9
Guinea	4.0	47	25	2.3	31	216	44	5.7
Ivory Coast	4.4	46	23	2.4	29	138	43	6.4
Liberia	1.2	41	23	1.9	37	188	37	1.6
Mali	5.2	50	25	2.4	29	120	46	7.6
Mauritania	1.2	45	23	2.2	32	187	—	1.7
Niger	4.0	52	23	2.9	24	200	46	6.2
Nigeria	56.5	50	25	2.6	27	—	43	84.7
Senegal	4.0	46	22	2.4	29	—	42	5.8
Sierra Leone	2.7	45	22	2.3	31	136	—	3.9
Togo	1.9	51	24	2.6	27	127	48	2.8
Upper Volta	5.5	49	28	2.1	33	182	42	7.7
EASTERN AFRICA	**100**	**47**	**21**	**2.6**	**27**	**—**	**44**	**149**
Burundi	3.7	48	25	2.3	31	150	47	5.3
Ethiopia	25.6	46	25	2.1	33	—	—	35.7
Kenya	11.2	50	20	3.1	23	—	46	17.9
Malagasy Republic	7.1	46	22	2.7	26	102	46	10.8
Malawi	4.6	49	25	2.5	28	148	45	6.8
Mauritius	0.9	27	8	1.9	37	72	41	1.2
Mozambique*	7.9	43	23	2.1	33	—	—	11.1
Reunion*	0.5	37	9	3.1	23	—	—	0.7
Rwanda	3.7	52	23	2.9	24	137	—	5.7
Somalia	2.9	46	24	2.4	29	—	—	4.2
Southern Rhodesia	5.2	48	14	3.4	21	122	47	8.6
Tanzania (United Republic of)	13.6	47	22	2.6	27	162	42	20.3
Uganda	8.8[4]	43	18	2.6	27	160	41	13.1
Zambia	4.4	50	20	3.0	24	259	45	7.0
MIDDLE AFRICA	**37**	**46**	**23**	**2.2**	**32**	**—**	**42**	**52**
Angola*	5.8	50	30	2.1	33	—	42	8.1
Cameroon (West)	5.9	43	21	2.2	32	137	39	8.4
Central African Republic	1.6	48	26	2.2	32	190	42	2.2
Chad	3.8	48	23	2.4	29	160	46	5.5
Congo (Dem. Republic)	17.8	44	21	2.3	31	104	42	25.8
Congo (Republic of)	1.0	44	23	2.3	31	180	—	1.4
Equatorial Guinea	0.3	35	22	1.4	50	—	—	0.4
Gabon	0.5	35	26	0.9	78	229	36	0.6
SOUTHERN AFRICA	**23**	**41**	**17**	**2.4**	**29**	**—**	**40**	**34**
Botswana	0.6	44	23	2.2	32	—	43	0.9
Lesotho	1.1	40	23	1.8	39	181	43	1.4
South Africa	20.6	40	16	2.4	29	—	40	29.7
Namibia (Southwest Africa)*	0.6	44	25	2.0	35	—	40	0.9
Swaziland	0.4	52	22	3.0	24	—	—	0.7

Region or Country	Population Estimates Mid-1971 (millions) †	Annual Births per 1,000 Population ‡	Annual Deaths per 1,000 Population ‡	Annual Rate of Population Growth (percent) ○	Number of Years to Double Population □	Annual Infant Mortality (Deaths under one year per 1,000 Live Births) ‡	Population under 15 Years (percent) ▶	Population Projections to 1985 (millions) †
ASIA	**2,104[2]**	**38**	**15**	**2.3**	**31**	**—**	**40**	**2,874**
SOUTHWEST ASIA	**79**	**44**	**15**	**2.9**	**24**	**—**	**43**	**121**
Cyprus	0.6	23	8	0.9	78	27	35	0.7
Iraq	10.0	49	15	3.4	21	—	45	16.7
Israel	3.0	26	7	2.4	29	23	33	4.0
Jordan	2.4	48	16	3.3	21	—	46	3.9
Kuwait	0.8	43	7	8.2	9	36	38	2.4
Lebanon	2.9	—	—	3.0	24	—	—	4.3
Muscat and Oman	0.7	42	11	3.1	23	—	—	1.1
Saudi Arabia	8.0	50	23	2.8	25	—	—	12.2
Southern Yemen	1.3	—	—	2.8	25	—	—	2.0
Syria	6.4	47	15	3.3	21	—	46	10.5
Turkey	36.5	43	16	2.7	26	155	44	52.8
Yemen (Arab Republic)	5.9	50	23	2.8	25	—	—	9.1
MIDDLE SOUTH ASIA	**783**	**44**	**16**	**2.7**	**26**	**—**	**43**	**1,137**
Afghanistan	17.4	50	26	2.5	28	—	—	25.0
Bhutan	0.9	—	—	2.2	32	—	—	1.2
Ceylon	12.9	32	8	2.4	29	48	41	17.7
India	569.5[4]	42	17	2.6	27	139	41	807.6
Iran	29.2	48	18	3.0	24	—	46	45.0
Nepal	11.5	45	23	2.2	32	—	40	15.8
Pakistan	141.6	50	18	3.3	21	142	45	224.2
SOUTHEAST ASIA	**295**	**43**	**15**	**2.8**	**25**	**—**	**44**	**434**
Burma	28.4	40	17	2.3	31	—	40	39.2
Cambodia	7.3	45	16	3.0	24	127	44	11.3
Indonesia	124.9	47	19	2.9	24	125	42	183.8
Laos	3.1	42	17	2.5	28	—	—	4.4
Malaysia	11.1	37	8	2.8	25	—	44	16.4
Philippines	39.4	46	12	3.4	21	72	47	64.0
Singapore	2.2	25	5	2.4	29	—	43	3.0
Thailand	37.4	42	10	3.3	21	—	43	57.7
Vietnam (Dem. Republic of)	21.6	—	—	2.1	33	—	—	28.2
Vietnam (Republic of)	18.3	—	—	2.1	33	—	—	23.9
EAST ASIA	**946**	**30**	**13**	**1.8**	**39**	**—**	**36**	**1,182**
China (Mainland)	772.9	33	15	1.8	39	—	—	964.6
China (Taiwan)	14.3	26	5	2.3	31	19	44	19.4
Hong Kong*	4.3	21	5	2.5	28	21	40	6.0
Japan	104.7	18	7	1.1	63	15	25	121.3
Korea (Dem. People's Rep. of)	14.3	39	11	2.8	25	—	—	20.7
Korea (Republic of)	32.9	36	11	2.5	28	—	42	45.9
Mongolia	1.3	42	10	3.1	23	—	44	2.0
Ryukyu Islands*	1.0	22	5	1.7	41	11	39	1.3
NORTHERN AMERICA	**229[2]**	**18**	**9**	**1.2**	**58**	**—**	**30**	**280**
Canada	21.8	17.6	7.3	1.7	41	20.8	33	27.3
United States[3]	207.1	18.2	9.3	1.1	63	19.8	30	241.7
LATIN AMERICA	**291[2]**	**38**	**9**	**2.9**	**24**	**—**	**42**	**435**
MIDDLE AMERICA	**70**	**43**	**9**	**3.4**	**21**	**—**	**46**	**112**
Costa Rica	1.9	45	8	3.8	19	60	48	3.2
El Salvador	3.6	47	13	3.4	21	63	45	5.9
Guatemala	5.3	42	13	2.9	24	94	46	7.9
Honduras	2.8	49	16	3.4	21	—	51	4.6
Mexico	52.5[4]	42	9	3.4	21	66	46	84.4
Nicaragua	2.1	46	16	3.0	24	—	48	3.3
Panama	1.5	41	8	3.3	21	41	43	2.5

Region or Country	Population Estimates Mid-1971 (millions) †	Annual Births per 1,000 Population ‡	Annual Deaths per 1,000 Population ‡	Annual Rate of Population Growth (percent) ○	Number of Years to Double Population □	Annual Infant Mortality (Deaths under one year per 1,000 Live Births) ‡	Population under 15 Years (percent) ▲	Population Projections to 1985 (millions) †
WORLD	**3,706**[1]	**34**	**14**	**2.0**	**35**	**—**	**37**	**4,933**
CARIBBEAN	**26**	**34**	**10**	**2.2**	**32**	**—**	**40**	**36**
Barbados	0.3	21	8	0.8	88	42	38	0.3
Cuba	8.6	27	8	1.9	37	40	37	11.0
Dominican Republic	4.4[4]	48	15	3.4	21	64	47	7.3
Guadeloupe*	0.4	32	8	2.4	29	35	42	0.5
Haiti	5.4	44	20	2.5	28	—	42	7.9
Jamaica	2.0	33	8	2.1	33	39	41	2.6
Martinique*	0.4	30	8	1.9	37	34	42	0.5
Puerto Rico*	2.9	24	6	1.4	50	29	39	3.4
Trinidad & Tobago	1.1	30	7	1.8	39	37	43	1.3
TROPICAL SOUTH AMERICA	**155**	**39**	**9**	**3.0**	**24**	**—**	**43**	**236**
Bolivia	4.8	44	19	2.4	29	—	44	6.8
Brazil	95.7	38	10	2.8	25	170	43	142.6
Colombia	22.1	44	11	3.4	21	78	47	35.6
Ecuador	6.3	45	11	3.4	21	86	48	10.1
Guyana	0.8	37	8	2.9	24	40	46	1.1
Peru	14.0	43	11	3.1	23	62	45	21.6
Surinam*	0.4	41	7	3.2	22	30	46	0.6
Venezuela	11.1	41	8	3.4	21	46	46	17.4
TEMPERATE SOUTH AMERICA	**40**	**26**	**9**	**1.8**	**39**	**—**	**33**	**51**
Argentina	24.7	22	9	1.5	47	58	29	29.6
Chile	10.0[4]	34	11	2.3	31	92	40	13.6
Paraguay	2.5	45	11	3.4	21	52	45	4.1
Uruguay	2.9	21	9	1.2	58	50	28	3.4
EUROPE	**466**[2]	**18**	**10**	**0.8**	**88**	**—**	**25**	**515**
NORTHERN EUROPE	**81**	**16**	**11**	**0.6**	**117**	**—**	**24**	**90**
Denmark	5.0	14.6	9.8	0.5	140	14.8	24	5.5
Finland	4.7	14.5	9.8	0.4	175	13.9	27	5.0
Iceland	0.2	20.7	7.2	1.2	58	11.7	34	0.3
Ireland	3.0	21.5	11.5	0.7	100	20.6	31	3.5
Norway	3.9	17.6	9.9	0.9	78	13.7	25	4.5
Sweden	8.1	13.5	10.4	0.5	140	13.0	21	8.8
United Kingdom	56.3	16.6	11.9	0.5	140	18.6	23	61.8
WESTERN EUROPE	**150**	**16**	**11**	**0.6**	**117**	**—**	**24**	**163**
Austria	7.5	16.5	13.4	0.4	175	25.4	24	8.0
Belgium	9.7	14.6	12.4	0.4	175	21.8	24	10.4
France	51.5	16.7	11.3	0.7	100	16.4	25	57.6
Germany (Federal Republic of)	58.9	15.0	12.0	0.4	175	23.3	23	62.3
Luxembourg	0.4	13.5	12.6	1.0	70	16.7	22	0.4
Netherlands	13.1	19.2	8.4	1.1	63	13.2	28	15.3
Switzerland	6.4	16.5	9.3	1.1	63	15.4	23	7.4
EASTERN EUROPE	**105**	**17**	**10**	**0.8**	**88**	**—**	**25**	**116**
Bulgaria	8.6	17.0	9.5	0.7	100	30.5	24	9.4
Czechoslovakia	14.8	15.5	11.2	0.5	140	22.9	25	16.2
Germany (Dem. Republic)	16.2	14.0	14.3	0.1	700	20.1	22	16.9
Hungary	10.3	15.0	11.3	0.4	175	35.7	23	11.0
Poland	33.3	16.3	8.1	0.9	78	34.3	30	38.2
Romania	20.6	23.3	10.1	1.3	54	54.9	26	23.3

Region or Country	Population Estimates Mid-1971 (millions) †	Annual Births per 1,000 Population ‡	Annual Deaths per 1,000 Population ‡	Annual Rate of Population Growth (percent) ○	Number of Years to Double Population □	Annual Infant Mortality (Deaths under one year per 1,000 Live Births) ‡	Population under 15 Years (percent) ▶	Population Projections to 1985 (millions) †
SOUTHERN EUROPE	**130**	**19**	**9**	**0.9**	**78**	—	**27**	**146**
Albania	2.2	35.6	8.0	2.7	26	86.8	—	3.3
Greece	9.0	17.4	8.2	0.8	88	31.9	25	9.7
Italy	54.1	17.6	10.1	0.8	88	30.3	24	60.0
Malta	0.3	15.8	9.4	−0.8	—	24.3	32	0.3
Portugal	9.6	19.8	10.6	0.7	100	56.8	29	10.7
Spain	33.6	20.2	9.2	1.0	70	29.8	27	38.1
Yugoslavia	20.8	18.8	9.2	1.0	70	56.3	30	23.8
USSR	**245**	**17.0**	**8.1**	**1.0**	**70**	**25.7**	**28**	**286.9**
OCEANIA	**20**[2]	**25**	**10**	**2.0**	**35**	—	**32**	**27**
Australia	12.8	20.0	9.1	1.9	37	17.7	29	17.0
Fiji	0.5	29	5	2.7	26	22	45	0.8
New Zealand	2.9	22.5	8.7	1.7	41	16.9	33	3.8

WORLD AND REGIONAL POPULATION (millions)

	WORLD	ASIA	EUROPE	USSR	AFRICA	NORTH AMERICA	LATIN AMERICA	OCEANIA
MID-1971	**3706**	**2104**	**466**	**245**	**354**	**229**	**291**	**20**
UN MEDIUM ESTIMATE, 2000	**6494**	**3777**	**568**	**330**	**818**	**333**	**652**	**35**

FOOTNOTES

† Estimates from United Nations. *"Total Population Estimates for World, Regions and Countries, Each Year, 1950-1985,"* Population Division Working Paper No. 34, October 1970.

‡ Latest available year. Except for Northern American rates, estimates are essentially those available as of January 1971 in UN *Population and Vital Statistics Report*. Series A, Vol. XXIII, No. 1, with adjustments as deemed necessary in view of deficiency of registration in some countries.

▶ Latest available year. Derived from UN *World Population Prospects, 1965-85, As Assessed in 1968,* Population Division Working Paper No. 30, December 1969 and UN *Demographic Yearbook, 1967.*

§ 1968 data supplied by the International Bank for Reconstruction and Development.

○ Annual rate of population growth (composed of the rate of natural increase modified by the net rate of in- or out-migration) is derived from the latest available published estimates by the United Nations, except where substantiated changes have occurred in birth rates, death rates or migration streams.

□ Assuming no change in growth rate.

* Nonsovereign country.

[1] Total reflects UN adjustments for discrepancies in international migration data.

[2] Regional population totals take into account small areas not listed on the *Data Sheet.*

[3] US figures are based on Series D projections of the 1970 census and vital statistics data available as of April 1971.

[4] In these countries, the UN estimates show a variation of more than 3 percent from recent census figures. Because of uncertainty as to the completeness or accuracy of census data, the UN estimates are used.

NOTE: The completeness and accuracy of data in many developing countries are subject to deficiencies of varying degree. In some cases, the data shown are estimates prepared by the United Nations.

The Population Reference Bureau, Inc. was founded in 1929 to educate the public about the implications of population growth and other demographic trends. With the advice of demographers, ecologists, economists, sociologists, political scientists and other scholars, the Bureau issues Population Bulletins, Profiles, Selections, *ancillary textbooks, an annual* World Population Data Sheet *and other publications. A list of publications is available from the Population Reference Bureau, 1755 Massachusetts Avenue, N.W., Washington, D. C. 20036. Regular Membership $8.00; Teacher or Student Membership $5.00; Library Subscription $5.00.*

Index

L

M

N

O

P

R

S